A-Level Maths for OCR

Paul Sanders

Published in 2005 by:
Nelson Thornes Ltd
Delta Place
27 Bath Road
CHELTENHAM
GL53 7TH
United Kingdom

08 09 / 10 9 8 7 6 5 4 3 2

A catalogue record for this book is available from the British Library

ISBN 978 0 7487 9453 9

Sample paper written by Val Dixon

Page make-up by Mathematical Composition Setters Ltd, Salisbury, United Kingdom

Printed and bound in Spain by Graphycems

Acknowledgements

We are grateful to the Oxford Cambridge and RSA Examination Board for permission to reproduce all the questions marked OCR.
All answers provided for examination questions are the sole responsibility of the author.

The publishers have made every effort to contact copyright holders but apologise if any have been overlooked.

CONTENTS

Possible Teaching Orders for C1 **(inside front cover)**

Introduction and Use of Calculators **vi**

1. Indices **1**

Basic rules of indices **1**
Further rules of indices **3**
Equations with indices **7**

2. Polynomials **11**

Function notation **11**
Polynomials **12**
Arithmetic of polynomials **13**
Identities **14**
Algebra with polynomials **16**
Factorisation of simple polynomials **18**

3. Quadratic Equations **27**

Solving simple quadratic equations **27**
Solving quadratic equations by factorisation **28**
Completing the square **30**
The formula for solving quadratic equations **34**
Using quadratic equations to solve problems **36**

4. Co-ordinate Geometry: Straight Lines **42**

Introductory ideas **42**
The equations of straight lines **48**
Parallel and perpendicular lines **53**
Using simultaneous equations to find points of intersection **56**
Areas of shapes **58**

5. Surds **64**

Simplifying surd expressions **64**
Division of an expression by $a + b\sqrt{x}$ **67**

6. The Quadratic Function **72**

Sketching the graph of a quadratic function **72**
Sketching the graph of a quadratic function that can be factorised **74**
The completed square format for $ax^2 + bx + c$ **77**
Sketching the graph of a quadratic function that cannot be factorised **79**
Linear and quadratic inequalities **81**
The discriminant of the equation $ax^2 + bx + c = 0$ **85**

7. The Gradient of a Curve 89

Tangents and gradients 89
The small chords method for estimating the gradient of a curve 90
Using a spreadsheet to investigate the gradient of $y = x^n$ 93
Practical differentiation 1 96
Using a spreadsheet to investigate the gradient of $f(x) + g(x)$ 99
Practical differentiation 2 101

8. Further Quadratic Equations 108

Equations that are reducible to quadratic equations 108
Simultaneous equations: one linear, one quadratic 110
Using the discriminant 112

9. Applications of Differentiation 115

Rates of change 115
Increasing and decreasing functions 116
Stationary points 117
Optimisation problems 121
Tangents and normals 125
The second derivative 128
Using the second derivative to determine the nature of a stationary point 129

10. Co-ordinate Geometry: The Circle 137

The equation of a circle 137
The tangent to a circle 138
Geometrical properties of circles 140
Intersection of a line with a circle 143

11. Graphs and Transformations 148

A catalogue of commonly occurring graphs 148
Sketching the graph of polynomials given in factorised form 151
The effect of transformations on graphs 156

Revision Exercises 170

Indices and surds 170
(revise Chapters 1 and 5 before attempting this exercise)

Polynomials 172
(revise Chapters 2, 3, 6 and 8 before attempting this exercise)

Co-ordinate geometry and graphs 175
(revise Chapters 4, 10 and 11 before attempting this exercise)

Differentiation 179
(revise Chapters 7 and 9 before attempting this exercise)

Sample Paper 181

Answers 183

Index 199

Module C1 Formulae (inside back cover)

INTRODUCTION

A-Level Maths for OCR is a brand new series from Nelson Thornes designed to give you the best chance of success in Advanced Level Maths. This book fully covers the OCR **C1** module specification.

In each chapter, you will find a number of key features:

- A beginning of chapter **OBJECTIVES** section, so you can see clearly what you should learn from each chapter
- **WORKED EXAMPLES** taking you through common questions, step by step
- Carefully graded **EXERCISES** to give you thorough practice in all concepts and skills
- Highlighted **KEY POINTS** to help you see at a glance what you need to know for the exam
- An **IT ICON** (IT) to highlight areas where IT software such as Excel may be used
- **EXTENSION** boxes with background information and additional theory
- An end-of-chapter **SUMMARY** to help with your revision
- An end-of-chapter **REVISION EXERCISE** so you can test your understanding of the chapter

At the end of the book, you will find a **MODULE REVISION EXERCISE** containing exam-type questions for the entire module. This is divided into four sections, mirroring the structure of the specification. [Each section tells you which chapters you should have done.]

Finally, there is a **SAMPLE EXAM PAPER** written by an OCR examiner which you can do under timed exam conditions to see just how well prepared you are for the real exam.

USE OF CALCULATORS

Calculators are **not** allowed in the examination for the C1 module.

When you see this symbol next to a question in the book, you will probably need to use a calculator to answer the question.

You should be able to answer all other questions without using a calculator.

1 Indices

The purpose of this chapter is to enable you to

- understand indices and use the laws of indices

You will already know that

3^5 is shorthand for $3 \times 3 \times 3 \times 3 \times 3 = 243$ and 7^4 is shorthand for $7 \times 7 \times 7 \times 7 = 2401$.

Similarly,

a^5 is shorthand for $a \times a \times a \times a \times a$ and y^4 is shorthand for $y \times y \times y \times y$.

In the expression 7^4, 4 is called the **index** or **power** and 7 is called the **base**. Similarly, in the expression a^5, 5 is the index and a is the base.

Basic Rules of Indices

Multiplication

It is possible to multiply two expressions involving powers of the same number by recalling what the index notation means:

$$3^2 \times 3^5 = (3 \times 3) \times (3 \times 3 \times 3 \times 3 \times 3) = 3 \times 3 \times 3 \times 3 \times 3 \times 3 \times 3 = 3^7$$

and

$$a^3 \times a^2 = (a \times a \times a) \times (a \times a) = a \times a \times a \times a \times a = a^5.$$

If you notice that

$$3^2 \times 3^5 = 3^7 = 3^{2+5} \quad \text{and} \quad a^3 \times a^2 = a^5 = a^{3+2}$$

then you can see that there is a much quicker way of multiplying together two powers of the same number:

Rule 1

$$a^m \times a^n = a^{m+n}$$

Raising a Power to a Power

Consider the two examples

Using rule 1

$$(3^2)^4 = 3^2 \times 3^2 \times 3^2 \times 3^2 = 3^{2+2+2+2} = 3^8 = 3^{2\times 4}$$

and

Using rule 1

$$(c^3)^2 = c^3 \times c^3 = c^{3+3} = c^6 = c^{3\times 2}.$$

Generalising these results gives:

Rule 2

$$(a^m)^p = a^{mp}$$

Division

It is possible to divide powers of the same number by writing the powers out longhand and then cancelling the common factors of the numerator and denominator:

$$3^6 \div 3^2 = \frac{3 \times 3 \times 3 \times 3 \times \cancel{3} \times \cancel{3}}{\cancel{3} \times \cancel{3}} = 3 \times 3 \times 3 \times 3 = 3^4$$

and

$$a^7 \div a^2 = \frac{a \times a \times a \times a \times a \times \cancel{a} \times \cancel{a}}{\cancel{a} \times \cancel{a}} = a \times a \times a \times a \times a = a^5.$$

Noticing that

$$3^5 \div 3^2 = 3^3 = 3^{5-2} \quad \text{and} \quad a^7 \div a^3 = a^4 = a^{7-3}$$

you can see that, in general:

Rule 3

$$a^m \div a^n = a^{m-n}$$

These three rules can be used to simplify expressions. For example:

$4^7 \times 4^5 = 4^{7+5} = 4^{12}$ (using rule 1)

$(8^3)^5 = 8^{3\times5} = 8^{15}$ (using rule 2)

$p^3 \times p^5 = p^8$ (using rule 1)

$(r^4)^3 = r^{4\times3} = r^{12}$ (using rule 2)

$\frac{7^{11}}{7^8} = 7^{11} \div 7^8 = 7^{11-8} = 7^3$ (using rule 3)

$f^7 \div f^5 = f^{7-5} = f^2$ (using rule 3)

$p^6 \div p = p^6 \div p^1 = p^{6-1} = p^5$ (using rule 3)

Remember that p is the same as p^1

$3p^2 \times 5p^4 = 3 \times 5 \times p^2 \times p^4 = 15p^6$ (using rule 1)

$100t^5 \div 5t^2 = \frac{100t^5}{5t^2} = \frac{100}{5} \times \frac{t^5}{t^2} = 20t^3$ (using rule 3)

Remember that these rules can only be used when the powers involved are powers of the same base. The expression $p^4 \times q^{11}$ can only be simplified to p^4q^{11} or $q^{11}p^4$.

Expressions such as $4p^3q^2 \times 8p^4q^2$ can be simplified through two usages of rule 1:

$$4p^3q^2 \times 8p^4q^2 = (4 \times 8) \times (p^3 \times p^4) \times (q^2 \times q^2) = 32 \times p^7 \times q^4 = 32p^7q^4.$$

Similarly, two usages of rule 3 gives

$$20p^5q^8 \div 4p^2q^4 = \frac{20p^5q^8}{4p^2q^4} = \frac{20}{4} \times \frac{p^5}{p^2} \times \frac{q^8}{q^4} = 5 \times p^3 \times q^4 = 5p^3q^4.$$

EXERCISE 1

1 Find the values of

i) 2^4 **ii)** 3^5 **iii)** 4^3 **iv)** $(-2)^3$ **v)** $(-3)^4$ **vi)** $(1^2 + 3^2)^5$

2 Find, correct to two decimal places, the values of
i) 2.1^4 **ii)** 3.12^7 **iii)** 0.84^5 **iv)** $(-1.85)^5$ **v)** 17.63^3

3 Simplify the following expressions, leaving your answers in index form:
a) $2^4 \times 2^7$ **b)** $3^8 \div 3^2$ **c)** $(5^3)^7$
d) $2^3 \times 2^4 \times 2^6$ **e)** $(7^2)^4 \times 7$ **f)** $(5^3 \times 5^8) \div 5^4$
g) $(4^3)^5 \div 4^7$ **h)** $(5^6 \div 5)^4$ **i)** $\dfrac{3^6 \times 3^2 \times 3^8}{3^3 \times 3^4}$

4 Simplify the following expressions, leaving your answers in index form:
a) $(a^2)^8$ **b)** $y \times y^3$ **c)** $c^3 \times c^7 \times c^8$ **d)** $\dfrac{z^9}{z}$
e) $\dfrac{y^7 \times y}{y^4}$ **f)** $(y^3 \times y^5)^3$ **g)** $(t^9 \div t^4)^5$ **h)** $(t^2 \times t)^5$
i) $\dfrac{(p^7 \times p)^3}{(p^4 \div p)}$ **j)** $\dfrac{(x^2)^3}{x}$ **k)** $\dfrac{d^8 \times d^9}{(d^3)^5}$ **l)** $\dfrac{t^{12}}{(t^4 \times t)^2}$

5 Simplify the following expressions:
a) $4s^2t^5 \times 5s^3t^3$ **b)** $24s^8t^6 \div 8s^2t^5$ **c)** $36s^5t^3 \div 9s^4t$
d) $4s^2t^4u^2 \times 6st^2u^3$ **e)** $2s^2t^3 \times 5st^4 \times 3s^3$ **f)** $(5s^3t^2 \times 4st^4) \div 10st^5$

Further Rules of Indices

So far only indices which are positive integers have been considered. How are expressions such as p^0, p^{-3}, $p^{\frac{1}{3}}$ and $p^{\frac{2}{5}}$ to be interpreted?

Interpreting a^0

Since any non-zero number divided by itself gives an answer of 1, you know that $3^4 \div 3^4 = 1$.

However, rule 3 gives $3^4 \div 3^4 = 3^{4-4} = 3^0$.

Putting these two observations together leads to the conclusion that $3^0 = 1$.

This argument can be generalised to obtain

Rule 4

$$a^0 = 1$$

Interpreting a^{-n}

Consider $3^3 \div 3^5$. You know that it is possible to write

$$3^3 \div 3^5 = \frac{\cancel{3} \times \cancel{3} \times \cancel{3}}{3 \times 3 \times \cancel{3} \times \cancel{3} \times \cancel{3}} = \frac{1}{3 \times 3} = \frac{1}{3^2}$$

but, rule 3 suggests that

$$3^3 \div 3^5 = 3^{3-5} = 3^{-2}$$

so it must be that $3^{-2} = \dfrac{1}{3^2}$.

In general:

Rule 5

$$a^{-n} = \frac{1}{a^n}$$

It is easy to check that rules 1–3 still apply when working with negative indices.

For example, consider $p^{-7} \div p^{-2}$.

Rule 3 would suggest that $p^{-7} \div p^{-2} = p^{-7-(-2)} = p^{-5}$ and you know that

$$p^{-7} \div p^{-2} = \frac{1}{p^7} \div \frac{1}{p^2} = \frac{1}{p^7} \times \frac{p^2}{1} = \frac{p \times p}{p \times p \times p \times p \times p \times p \times p}$$
$$= \frac{1}{p^5} = p^{-5}.$$

Interpreting $a^{\frac{1}{n}}$ When n is Odd

Start by trying to interpret the expression $a^{\frac{1}{3}}$.

Consider $5^{\frac{1}{3}}$.

Using rule 2, you would expect $(5^{\frac{1}{3}})^3 = 5^{3 \times \frac{1}{3}} = 5^1 = 5$ so you require $5^{\frac{1}{3}}$ to be a solution of the equation $x^3 = 5$. This equation has just **one** solution, namely $x = \sqrt[3]{5}$.

You now know that $5^{\frac{1}{3}} = \sqrt[3]{5}$ = the cube root of 5.

An identical argument leads to $(-27)^{\frac{1}{3}} = \sqrt[3]{-27} = -3$.

In general:

If a is any number then $a^{\frac{1}{3}} = \sqrt[3]{a}$

This argument can be generalised for any **odd** number, n:

For an odd number, n:
if a is any number then $a^{\frac{1}{n}} = \sqrt[n]{a}$

Interpreting $a^{\frac{1}{n}}$ When n is Even

Now consider the interpretation of $a^{\frac{1}{2}}$.

Consider $3^{\frac{1}{2}}$.

Using rule 2, you would expect $(3^{\frac{1}{2}})^2 = 3^{2 \times \frac{1}{2}} = 3^1 = 3$
so $3^{\frac{1}{2}}$ should be a solution of the equation $x^2 = 3$.

This equation has **two** solutions: $\sqrt{3}$ and $-\sqrt{3}$.

Remember that $\sqrt{x}$ means the **positive** number whose square is x.

The value of $3^{\frac{1}{2}}$ is **defined** to be the **positive** square root of 3: i.e. $3^{\frac{1}{2}} = \sqrt{3}$.

Notice that it is not possible find a value for $(-5)^{\frac{1}{2}}$ since the equation $x^2 = -5$ has no real solutions.

In general, it can be said that:

If $a \geqslant 0$ then $a^{\frac{1}{2}} = \sqrt{a}$
If $a < 0$ then $a^{\frac{1}{2}}$ does not exist.

This argument can be generalised for any **even** number n to give the result

For an even number, n:

If $a \geqslant 0$ then $a^{\frac{1}{n}} = \sqrt[n]{a}$
If $a < 0$ then $a^{\frac{1}{n}}$ does not exist.

Putting the results for odd and even values of n together, we have

Rule 6

$$a^{\frac{1}{n}} = \sqrt[n]{a}$$

(unless n is even and a is negative in which case $a^{\frac{1}{n}}$ does not exist).

Interpreting $a^{\frac{m}{n}}$

Now consider $27^{\frac{2}{3}}$.

$$27^{\frac{2}{3}} = (27^{\frac{1}{3}})^2 \qquad \text{(rule 2)}$$

$$\Rightarrow \quad 27^{\frac{2}{3}} = (\sqrt[3]{27})^2 \qquad \text{(rule 6)}$$

$$\Rightarrow \quad 27^{\frac{2}{3}} = (3)^2 = 9.$$

You could, instead, have written $27^{\frac{2}{3}} = (27^2)^{\frac{1}{3}} = (729)^{\frac{1}{3}} = \sqrt[3]{729} = 9$.

Similarly,

$$(-32)^{\frac{3}{5}} = ((-32)^{\frac{1}{5}})^3 = (\sqrt[5]{-32})^3 = (-2)^3 = -8.$$

Again, you could, instead, have written

$$(-32)^{\frac{3}{5}} = ((-32)^3)^{\frac{1}{5}} = (-32\,768)^{\frac{1}{5}} = \sqrt[5]{-32\,768} = -8.$$

In general:

Rule 7

$$a^{\frac{m}{n}} = (a^{\frac{1}{n}})^m = (\sqrt[n]{a})^m$$

or

$$a^{\frac{m}{n}} = (a^m)^{\frac{1}{n}} = (\sqrt[n]{a^m})$$

As a general rule you will find the relationship

$$a^{\frac{m}{n}} = (\sqrt[n]{a})^m$$

more useful in calculations since it generally keeps the numbers under consideration closer to 1 and thus makes the mental arithmetic easier.

Powers of Products and Quotients

The final two rules for working with indices will enable you to simplify expressions such as $(xy)^n$ and $\left(\frac{x}{y}\right)^n$.

The results $(pq)^3 = pq \times pq \times pq = p^3q^3$ and $\left(\frac{p}{q}\right)^4 = \frac{p}{q} \times \frac{p}{q} \times \frac{p}{q} \times \frac{p}{q} = \frac{p^4}{q^4}$

can be generalised to:

Rule 8

$$(pq)^n = p^n q^n$$

and

Rule 9

$$\left(\frac{p}{q}\right)^n = \frac{p^n}{q^n}$$

Evaluating or Simplifying Expressions Involving Indices

The nine rules of indices can be used to evaluate or simplify a wide range of expressions.

EXAMPLE 1

Without using a calculator, evaluate

a) 7^{-2} **b)** $4^{\frac{5}{2}}$ **c)** $(-64)^{-\frac{2}{3}}$ **d)** $(\frac{2}{3})^{-2}$

SOLUTION

a) $7^{-2} = \dfrac{1}{7^2} = \dfrac{1}{49}$

b) $4^{\frac{5}{2}} = (4^{\frac{1}{2}})^5 = (\sqrt{4})^5 = 2^5 = 32$

c) $(-64)^{-\frac{2}{3}} = \dfrac{1}{(-64)^{\frac{2}{3}}} = \dfrac{1}{((-64)^{\frac{1}{3}})^2} = \dfrac{1}{(\sqrt[3]{-64})^2} = \dfrac{1}{(-4)^2} = \dfrac{1}{16}$

d) $\left(\dfrac{2}{3}\right)^{-2} = \dfrac{1}{\left(\dfrac{2}{3}\right)^2} = \dfrac{1}{\left(\dfrac{4}{9}\right)} = \dfrac{9}{4}$

EXAMPLE 2

Simplify the following expressions:

a) $(5x^2)^3$ **b)** $\left(\dfrac{4}{a}\right)^3$ **c)** $\dfrac{(3\sqrt{x})^4}{18x^3}$ **d)** $(5\sqrt[4]{x})^2$

SOLUTION

a) $(5x^2)^3 = 5^3(x^2)^3 = 125x^6$ (using rules 8 and 2)

b) $\left(\dfrac{4}{a}\right)^3 = \dfrac{4^3}{a^3} = \dfrac{64}{a^3}$ (using rule 9)

Remember that $\sqrt{x} = x^{\frac{1}{2}}$.

c) $\dfrac{(3\sqrt{x})^4}{18x^3} = \dfrac{(3x^{\frac{1}{2}})^4}{18x^3} = \dfrac{3^4(x^{\frac{1}{2}})^4}{18x^3} = \dfrac{81x^2}{18x^3} = \dfrac{9}{2x}$

(using rules 8 and 2 and then cancelling out the common factor of $9x^2$)

d) $(5\sqrt[4]{x})^2 = (5x^{\frac{1}{4}})^2 = 5^2(x^{\frac{1}{4}})^2 = 25x^{\frac{1}{2}}$ or $25\sqrt{x}$ (using rules 8 and 2)

EXERCISE 2

1 Find the values of

a) 2^{-3} **b)** $9^{\frac{1}{2}}$ **c)** $25^{-\frac{1}{2}}$ **d)** $36^{\frac{3}{2}}$ **e)** $8^{-\frac{1}{3}}$

f) $8^{\frac{2}{3}}$ **g)** $125^{\frac{2}{3}}$ **h)** $64^{-\frac{1}{3}}$ **i)** $49^{-\frac{3}{2}}$ **j)** $16^{-\frac{3}{4}}$

k) $(\frac{3}{5})^{-2}$ **l)** $(\frac{4}{9})^{\frac{3}{2}}$ **m)** $(\frac{8}{27})^{-\frac{2}{3}}$ **n)** $(\frac{3}{2})^{-3}$ **o)** $(\frac{25}{4})^{-\frac{3}{2}}$

2 Simplify the following expressions:

a) $(3x^2)^4$ **b)** $\left(\dfrac{2x}{y^2}\right)^3$ **c)** $(4p^2q)^3$ **d)** $\left(\dfrac{16p^4}{q^6}\right)^{\frac{1}{2}}$ **e)** $(4y)^{-2}$

f) $\dfrac{(4x^2)^3}{(6x)^4}$ **g)** $(25x^4)^{\frac{1}{2}}$ **h)** $(y^6)^{-3}$ **i)** $(27y^6)^{\frac{2}{3}}$ **j)** $(32p^{10})^{-\frac{1}{5}}$

3 Simplify the following expressions:

a) $\dfrac{x^{\frac{3}{2}}}{\sqrt{x}}$ **b)** $\dfrac{(\sqrt{x})^6}{x}$ **c)** $\dfrac{(2\sqrt{x})^3}{x}$ **d)** $\sqrt{p} \times p^{\frac{5}{2}}$

e) $(2\sqrt{y})^3 \times (\sqrt{y})^5$ **f)** $\dfrac{x^5}{(\sqrt{x})^4}$ **g)** $\dfrac{(x\sqrt{x})^3}{x^4}$ **h)** $\dfrac{(x^2)^3}{(x\sqrt{x})^4}$

4 Simplify the following expressions:

a) $\dfrac{(x^2)^{\frac{1}{3}}\sqrt{x}}{x^{\frac{1}{6}}}$ **b)** $\dfrac{(x^3)^{\frac{2}{5}}(2\sqrt{x})^3}{x^{\frac{7}{10}}}$ **c)** $((\sqrt{y})^3 \div y^{\frac{5}{4}})^8$

Equations with Indices

If $a^p = a^q$ then it can be concluded that $p = q$, provided a is not 0, 1 or −1.

This fact enables you to solve many equations which involve indices.

EXAMPLE 3

Find p if $5^p = 125$.

SOL

Since $125 = 5^3$ the equation can be rewritten as

$5^p = 5^3 \quad \Rightarrow \quad p = 3.$

EXAMPLE 4

Find r if $4^r = \frac{1}{16}$.

SOL

Since $\dfrac{1}{16} = \dfrac{1}{4^2} = 4^{-2}$ we have $4^r = 4^{-2} \quad \Rightarrow \quad r = -2.$.

EXAMPLE 5

Find y if $5^{3y+8} = 25$.

SOLUTION

Since $25 = 5^2$ we have

$$\begin{aligned} & 5^{3y+8} = 5^2 \\ \Rightarrow\quad & 3y + 8 = 2 \\ \Rightarrow\quad & 3y = -6 \\ \Rightarrow\quad & y = -2. \end{aligned}$$

EXAMPLE 6

Find s if $3^{s+2} = 27^{5-4s}$.

SOLUTION

Since $27 = 3^3$ we have

$$\begin{aligned} & 3^{s+2} = (3^3)^{5-4s} \\ \Rightarrow\quad & 3^{s+2} = 3^{15-12s} \quad \text{(using rule 2)} \\ \Rightarrow\quad & s + 2 = 15 - 12s \\ \Rightarrow\quad & 13s = 13 \\ \Rightarrow\quad & s = 1. \end{aligned}$$

Notice how the first stage in each of the last four examples has been to rewrite the equation in an equivalent form **with the same base on each side of the equation.**

Equations of the form $x^n = a$ can be solved by taking the nth root of each side and this can be written in terms of indices.

EXAMPLE 7

Find p if $p^5 = 69$.

SOLUTION

To find p you need to take the fifth root of each side of the equation:

$$p^5 = 69$$
$$\Rightarrow \quad (p^5)^{\frac{1}{5}} = 69^{\frac{1}{5}}$$
$$\Rightarrow \quad p^1 = 69^{0.2}$$
$$\Rightarrow \quad p = 2.332 \qquad \text{(3 d.p.).}$$

You know that the equation $x^2 = 9$ has two solutions: 3 and −3;
and that the equation $y^4 = 625$ also has two solutions: 5 and −5.

You need to remember that, if n is an even non-zero integer and k is a positive number then the equation $x^n = k$ will have two solutions: $x = \sqrt[n]{k} = k^{\frac{1}{n}}$ and $x = -\sqrt[n]{k} = -(k^{\frac{1}{n}})$.

EXAMPLE 8

Solve the equation $6y^4 = 894$.

SOLUTION

$$6y^4 = 894$$
$$\Rightarrow \quad y^4 = 149$$
$$\Rightarrow \quad y = 149^{\frac{1}{4}} \quad \text{or} \quad y = -(149^{\frac{1}{4}})$$
$$\Rightarrow \quad y = 3.494 \quad \text{or} \quad y = -3.494 \qquad \text{(3 d.p.).}$$

Equations with fractional or negative indices can also be solved in a similar way.

EXAMPLE 9

Solve the equation $y^{\frac{1}{4}} = 2$.

SOLUTION

$$y^{\frac{1}{4}} = 2$$
$$\Rightarrow \quad (y^{\frac{1}{4}})^4 = 2^4$$
$$\Rightarrow \quad y^1 = 16 \qquad \text{(rule 3)}$$
$$\Rightarrow \quad y = 16.$$

Raising each side of the equation to the power four.

EXAMPLE 10

Solve the equation $t^{\frac{3}{4}} = 27$.

SOLUTION

$$t^{\frac{3}{4}} = 27$$
$$\Rightarrow \quad (t^{\frac{3}{4}})^{\frac{4}{3}} = 27^{\frac{4}{3}}$$
$$\Rightarrow \quad t^1 = (27^{\frac{1}{3}})^4 = (\sqrt[3]{27})^4$$
$$\Rightarrow \quad t = 3^4 = 81.$$

Raising each side of the equation to the power $\frac{4}{3}$.

EXAMPLE 11

Solve the equation $w^{-3} = 64$.

SOLUTION

$$w^{-3} = 64$$

$$\Rightarrow \quad (w^{-3})^{-\frac{1}{3}} = 64^{-\frac{1}{3}}$$

Raising each side of the equation to the power $-\frac{1}{3}$.

$$\Rightarrow \quad w^1 = \frac{1}{64^{\frac{1}{3}}} = \frac{1}{\sqrt[3]{64}}$$

$$\Rightarrow \quad w = \tfrac{1}{4}.$$

EXERCISE 3

1 Solve the equations

a) $3^p = 81$ **b)** $4^y = 16$ **c)** $5^t = \frac{1}{125}$ **d)** $3^{t-2} = 27$

e) $36^x = 6$ **f)** $2^{5p-4} = 8^p$ **g)** $6^{t-3} = 36^{2t-3}$ **h)** $3^{2p-4} = (\frac{1}{9})^{6-3p}$

i) $64^{s+1} = 16^{2s}$ **j)** $4^{t-5} = (\frac{1}{2})^{t+1}$

2 Find the solutions, correct to three significant figures, of the equations

a) $p^3 = 72$ **b)** $y^{10} = 3124$ **c)** $4y^5 = 6231$

d) $8y^3 = 6.56$ **e)** $5y^8 = 2.152$

3 Find the solutions of the equations

a) $u^{\frac{1}{3}} = 4$ **b)** $v^{\frac{1}{4}} = \frac{1}{3}$ **c)** $s^{\frac{1}{5}} = -2$

d) $t^{\frac{3}{5}} = 8$ **e)** $z^{-3} = 8$ **f)** $y^{-\frac{3}{4}} = 27$

Having studied this chapter you should know how

- to use the rules of indices:

1. $a^m \times a^n = a^{m+n}$
2. $(a^m)^p = a^{mp}$
3. $a^m \div a^n = a^{m-n}$
4. $a^0 = 1$
5. $a^{-n} = \dfrac{1}{a^n}$
6. $a^{\frac{1}{n}} = \sqrt[n]{a}$
7. $a^{\frac{m}{n}} = (a^{\frac{1}{n}})^m = (\sqrt[n]{a})^m$
   $a^{\frac{m}{n}} = (a^m)^{\frac{1}{n}} = \sqrt[n]{a^m}$
8. $(pq)^n = p^n q^n$
9. $\left(\dfrac{p}{q}\right)^n = \dfrac{p^n}{q^n}$

- to solve simple equations involving indices

REVISION EXERCISE

1 Simplify

a) $p^3 \times p^5$ **b)** $(p^4)^7$ **c)** $\dfrac{p^8}{(p^3)^5}$

2 Find the values of

a) $(16)^{-\frac{1}{2}}$ **b)** 72^0 **c)** $32^{\frac{3}{5}}$

3 Simplify

a) $\dfrac{(2y^6)^3}{(4y^4)^2}$ **b)** $\sqrt[3]{27y^{12}}$ **c)** $(5y^6)^{-2}$

4 Simplify

a) $\dfrac{(\sqrt{x})^5 x^{\frac{1}{3}}}{(\sqrt[6]{x})^{11}}$ **b)** $\sqrt{(4x^2)^3 \div (8x^3)^2}$

5 Solve the equations

a) $6^t = \frac{1}{36}$ **b)** $7^{t+6} = 49^{2t}$ **c)** $9^{y+8} = 27^{y+5}$

6 Solve the equations, giving your answers to three significant figures,

a) $p^{12} = 75$ **b)** $4r^5 + 7 = 100$ **c)** $s^{-5} = 4$

7 **i)** Express $(\frac{3}{4})^{-2}$ as an exact fraction in its simplest form

ii) Simplify $\dfrac{(2\sqrt{x})^4}{8x}$

(OCR Jun 2001 P1)

8 Solve the equations

a) $t^{\frac{1}{3}} = 4$ **b)** $u^{-3} = 1000$ **c)** $v^{\frac{2}{5}} = 4$

9 Evaluate

a) $(1^3 + 2^3 + 3^3)^{\frac{1}{2}}$ **b)** $(\frac{3}{5})^{-2}$ **c)** $(16)^{-\frac{3}{4}}$ **d)** $(-8)^{\frac{5}{3}}$

10 Simplify

a) $(x^2\sqrt{x})^2$ **b)** $x^{\frac{7}{2}} \div \sqrt{x}$ **c)** $\dfrac{(\sqrt[3]{x})^9}{(\sqrt{x})^4}$ **d)** $\left(\dfrac{2x^2}{\sqrt[3]{x}}\right)^3$

11 It is given that $a^p = 5$ and $a^q = 9$. In each of the following cases, determine the numerical value of

i) a^{p+q} **ii)** $2a^{-p}$ **iii)** $a^{2p-\frac{1}{2}q}$

(OCR Mar 1998 P1)

12 If $z = 36t^{-2}$

a) find the value of z when $t = 3$

b) find the possible values of t if $z = \frac{4}{9}$

2 Polynomials

The purpose of this chapter is to enable you to

- use function notation
- add, subtract and multiply polynomials
- factorise polynomials by finding a common factor
- factorise quadratic polynomials

Function Notation

Letters, such as f, g and h, are used to represent functions.

For example, suppose a rule or function, f, takes a number, multiplies it by 3 and then adds 5. The rule f can be described either by writing

$$\text{f}: \quad x \longrightarrow 3x + 5$$

or by writing

$$\text{f}(x) = 3x + 5.$$

If the rule f is applied to the number 2 then the result will be 11. This can be written as

$$\text{f}: \quad 2 \longrightarrow 11$$

or as

$$\text{f}(2) = 11.$$

Read this as "the function f applied to 2 gives 11" or "f of 2 is 11".

If $\text{g}(x) = x^2$ is written as shorthand for the rule g: $x \longrightarrow x^2$,

then $\text{g}(3) = 3^2 = 9$
and $\text{g}(-2) = (-2)^2 = 4$.

Moreover $\text{g}(2x) = (2x)^2 = 4x^2$
and $\text{g}(x - 3) = (x - 3)^2$.

Similarly, if you write $\text{h}(x) = x^3 + 5x - 7$ as shorthand for the rule h: $x \longrightarrow x^3 + 5x - 7$,

then $\text{h}(2) = 2^3 + 5 \times 2 - 7 = 8 + 10 - 7 = 11$
and $\text{h}(-0.1) = (-0.1)^3 + 5 \times (-0.1) - 7 = -0.001 - 0.5 - 7 = -7.501$.

Moreover $\text{h}(3x) = (3x)^3 + 5 \times (3x) - 7 = 27x^3 + 15x - 7$.

Recall from chapter 1 $(3x)^3 = 3^3x^3 = 27x^3$

EXERCISE 1

1 If h: $x \longrightarrow x^2 + 1$
i) find the values of
a) h(3) **b)** h(−2) **c)** h(0.3) **d)** h(12)
ii) sketch the graph of $y = \text{h}(x)$

2 If m: $x \longrightarrow x^3$

i) find the values of

a) m(4) **b)** m(–2) **c)** m(10) **d)** m(0.6)

ii) write down algebraic expressions for

a) m($2x$) **b)** m($x + 2$)

iii) find the values of α and β if m(α) = 125 and m(β) = –0.008

3 If f: $x \longrightarrow 3x^2 + 5x - 2$

i) find the values of

a) f(1) **b)** f(–5) **c)** f(0.5)

ii) write down algebraic expressions for

a) f($4x$) **b)** f($2x - 1$)

Polynomials

Functions with rules of the form

$$\mathrm{e}(x) = x^2, \quad \mathrm{f}(x) = 2x^3 + 5x^2 - 7, \quad \mathrm{g}(x) = x^5 + x - 3 \quad \text{and} \quad \mathrm{h}(x) = 32x^{12} + 5x^{20}$$

are examples of polynomials.

On the other hand, functions with rules of the form

$$\mathrm{k}(x) = \frac{1}{x}, \quad \mathrm{m}(x) = 3x^2 + 5\sqrt{x} \quad \text{and} \quad \mathrm{n}(x) = 3x^{0.2} - 5x^{-0.4}$$

are **not** examples of polynomials.

A polynomial is the addition of a series of terms each of which is a multiple of a non-negative integer power of x.

The **degree** of a polynomial is the highest power that appears in the polynomial.
Thus, e is a polynomial of degree 2, f is a polynomial of degree 3, g is a polynomial of degree 5, and h is a polynomial of degree 20.

A **quadratic function** is a polynomial of degree 2 and a **cubic function** is a polynomial of degree 3. Thus, p(x) = $-5x^2 + 3x - 9$ is a quadratic function and q(x) = $2x^3 - 11x$ is a cubic function.

For the polynomial q(x) = $2x^3 - 11x$, you can say that "2 is the **coefficient** of x^3 and –11 is the **coefficient** of x".

For the polynomial p(x) = $-5x^2 + 3x - 9$, you can say that "–5 is the coefficient of x^2, 3 is the coefficient of x and –9 is the coefficient of x^0". You can also say that –9 is the **constant term** of the polynomial f.

Arithmetic of Polynomials

Addition and subtraction of polynomials are straightforward algebraic processes:

If $f(x) = 2x^3 + 5x^2 - 7$ and $g(x) = x^2 + x - 3$

then

$$\begin{aligned} f(x) + g(x) &= 2x^3 + 5x^2 - 7 + x^2 + x - 3 \\ &= 2x^3 + 6x^2 + x - 10 \\ f(x) - g(x) &= 2x^3 + 5x^2 - 7 - (x^2 + x - 3) \\ &= 2x^3 + 5x^2 - 7 - x^2 - x + 3 \\ &= 2x^3 + 4x^2 - x - 4. \end{aligned}$$

The like terms can now be added, or subtracted, as appropriate.

Take care with the signs here!

Multiplication of a polynomial by a constant number is also straightforward:

If $f(x) = 2x^3 + 5x^2 - 7$ and $g(x) = x^2 + x - 3$
then

$$3f(x) = 3(2x^3 + 5x^2 - 7) = 6x^3 + 15x^2 - 21$$

Just multiply each term of the polynomial by the constant.

and

$$\begin{aligned} 2f(x) + 3g(x) &= 2(2x^3 + 5x^2 - 7) + 3(x^2 + x - 3) \\ &= 4x^3 + 10x^2 - 14 + 3x^2 + 3x - 9 \\ &= 4x^3 + 13x^2 + 3x - 23. \end{aligned}$$

To multiply two polynomials, each term of the first polynomial must be multiplied by each term of the second polynomial:

If $f(x) = 2x^3 + 5x^2 - 7$ and $g(x) = x^2 + x - 3$
then

$$\begin{aligned} f(x)g(x) &= (2x^3 + 5x^2 - 7)(x^2 + x - 3) \\ &= 2x^3(x^2 + x - 3) + 5x^2(x^2 + x - 3) - 7(x^2 + x - 3) \\ &= 2x^5 + 2x^4 - 6x^3 + 5x^4 + 5x^3 - 15x^2 - 7x^2 - 7x + 21 \\ &= 2x^5 + 7x^4 - x^3 - 22x^2 - 7x + 21. \end{aligned}$$

A grid can also be used to present this calculation:

	x^2	$+x$	-3
$2x^3$	$2x^5$	$+2x^4$	$-6x^2$
$+5x^2$	$+5x^4$	$+5x^3$	$-15x^2$
-7	$-7x^2$	$-7x$	$+21$

so

$$\begin{aligned} f(x)g(x) &= (2x^3 + 5x^2 - 7)(x^2 + x - 3) \\ &= 2x^5 + 2x^4 - 6x^3 + 5x^4 + 5x^3 - 15x^2 - 7x^2 - 7x + 21 \\ &= 2x^5 + 7x^4 - x^3 - 22x^2 - 7x + 21. \end{aligned}$$

EXERCISE 2

In this exercise, f, g, h and j are the polynomials defined by

$f(x) = x^4 + 3x - 7$ $\qquad$ $g(x) = 2x^3 + x$

$h(x) = 3x^4 - 5x^3 + 7x + 1$ $\qquad$ $j(x) = -2x^3$

1 **a)** Write down the degree of each of these polynomials

b) **i)** Write down the coefficient of x in $f(x)$

ii) Write down the coefficient of x^4 in $h(x)$

iii) Write down the coefficient of x^3 in $j(x)$

iv) Write down the coefficient of x^3 in $f(x)$

2 Express, in simplified form,

a) $f(x) + g(x)$ **b)** $f(x) - g(x)$ **c)** $f(x)j(x)$

d) $2f(x) + h(x)$ **e)** $g(x) + j(x)$ **f)** $3f(x) - h(x)$

g) $f(x)g(x)$ **h)** $3g(x) + 2h(x)$ **i)** $g(x)h(x)$

3 **a)** Find the coefficient of x^7 in $f(x)g(x)$ **b)** Find the coefficient of x^5 in $f(x)g(x)$

4 If $p(x)$ is a polynomial of degree α and $q(x)$ is a polynomial of degree β use your answer to question 2 to make predictions about the degree of

i) $p(x)q(x)$ **ii)** $p(x) + q(x)$

Try to prove your predictions.

Identities

The symbol ≡ is sometimes used in a mathematical statement to emphasise the fact that the statement is an identity where the two quantities on either side of the symbol are **always the same**.

For example, it is possible to write

$$(x - 3)(2x - 1) \equiv 2x^2 - 7x + 3$$

because, whatever the value of x, the value of $(x - 3)(2x - 1)$ is the same as the value of $2x^2 - 7x + 3$. You can say that "$(x - 3)(2x - 1)$ is **identically equal** to $2x^2 - 7x + 3$".

The statement $(x + 3)^2 = x^2 + 6x + 9$ is always true, whatever the value of x, so you could write $(x + 3)^2 \equiv x^2 + 6x + 9$.

However, the statement $x + 3 = 7$ is only true when x is 4 and it would therefore be incorrect to write $x + 3 \equiv 7$.

Proving Identities

Algebraic identities can usually be proved by starting with one side of the identity and progressing in a steady logical fashion towards the other side.

For example, to prove the identity

$$(2 + x)^3 \equiv 8 + 12x + 6x^2 + x^3$$

we could write

$$\begin{aligned}(2 + x)^3 &\equiv (2 + x)(2 + x)(2 + x) \\ &\equiv (2 + x)(4 + 2x + 2x + x^2) \\ &\equiv (2 + x)(4 + 4x + x^2) \\ &\equiv 8 + 8x + 2x^2 + 4x + 4x^2 + x^3 \\ &\equiv 8 + 12x + 6x^2 + x^3.\end{aligned}$$

Identical Polynomials

Suppose that the quadratic polynomials $f(x)$ and $g(x)$ are given by

$f(x) = ax^2 + bx + c$ where a, b and c are constant numbers
and $g(x) = px^2 + qx + r$ where p, q and r are constant numbers.

If $f(x) \equiv g(x)$, so that the value of $f(x)$ is equal to the value of $g(x)$ for **every value** of x,

then $f(0) = g(0) \Rightarrow c = r$
and $f(1) = g(1) \Rightarrow a + b + c = p + q + r$

$$\Rightarrow a + b + r = p + q + r$$
$$\Rightarrow a + b = p + q$$
$$\Rightarrow b = p + q - a \quad [1]$$

We know already that $c = r$

and $f(-1) = g(-1) \Rightarrow a - b + c = p - q + r$

$$\Rightarrow a - (p + q - a) + r = p - q + r$$
$$\Rightarrow a - (p + q - a) = p - q$$
$$\Rightarrow 2a - p - q = p - q$$
$$\Rightarrow 2a = 2p$$
$$\Rightarrow a = p$$

We know already that $c = r$ and $b = p + q - a$.

and equation [1] now simplifies to

$$b = p + q - p$$
$$\Rightarrow b = q.$$

So, if two quadratic polynomials give identical values for all values of x then the coefficients of x^2 must be equal, the coefficients of x must be equal and the constant terms must be equal.

This result can be generalised to polynomials of any degree.

EXAMPLE 1

Find the values of α, β and γ if

$$2x^3 + 7x^2 + 5x - 2 \equiv (x - 2)(\alpha x^2 + 1) + \beta x^2 + \gamma x.$$

SOLUTION

Since

$$(x - 2)(\alpha x^2 + 1) + \beta x^2 + \gamma x \equiv \alpha x^3 + x - 2\alpha x^2 - 2 + \beta x^2 + \gamma x$$
$$\equiv \alpha x^3 + \beta x^2 - 2\alpha x^2 + x + \gamma x - 2$$
$$\equiv \alpha x^3 + (\beta - 2\alpha)x^2 + (1 + \gamma)x - 2$$

you require

$$2x^3 + 7x^2 + 5x - 2 \equiv \alpha x^3 + (\beta - 2\alpha)x^2 + (1 + \gamma)x - 2.$$

You know that two polynomials are identical only if their coefficients are the same, so equating the coefficients gives

$$2 = \alpha$$

Looking at coefficients of x^3.

$$7 = \beta - 2\alpha$$
$$\Rightarrow 7 = \beta - 4$$
$$\Rightarrow \beta = 11$$

Looking at coefficients of x^2.

$$5 = 1 + \gamma$$
$$\Rightarrow \gamma = 4.$$

Looking at coefficients of x.

Algebra with Polynomials

Working with polynomials will frequently require accurate usage of the algebraic techniques that you met prior to this course.

EXAMPLE 2

The polynomial f is defined by $f(x) = x^2 + 5x + 7$

a) find f(2)
b) find a simplified expression for $f(x^3)$
c) prove that $f(2x - 1) \equiv 4x^2 + 6x + 3$

SOLUTION

a) $f(2) = 2^2 + 5 \times 2 + 7 = 4 + 10 + 7 = 21$
b) $f(x^3) = (x^3)^2 + 5x^3 + 7 = x^6 + 5x^3 + 7$
c) $f(2x - 1) \equiv (2x - 1)^2 + 5(2x - 1) + 7$
$\equiv 4x^2 - 4x + 1 + 10x - 5 + 7$
$\equiv 4x^2 + 6x + 3$

$(2x - 1)^2$ means $(2x - 1)$ multiplied by itself so we can write

$$(2x - 1)^2 \equiv (2x - 1)(2x - 1) \equiv 4x^2 - 4x + 1.$$

EXAMPLE 3

Find integers p and q so that

$$x^2 + 8x - 17 \equiv (x + p)^2 + q.$$

SOLUTION

Since

$$(x + p)^2 \equiv (x + p)(x + p) \equiv x^2 + xp + px + p^2 \equiv x^2 + 2px + p^2$$

you know that

$$(x + p)^2 + q \equiv x^2 + 2px + p^2 + q$$

so you want

$$x^2 + 8x - 17 \equiv x^2 + 2px + p^2 + q.$$

From the previous section you know that two quadratic polynomials are identical only if their coefficients are the same.

Considering the coefficient of x:

$$8 = 2p \quad \Rightarrow \quad p = 4.$$

Considering the constant coefficient:

$$-17 = p^2 + q \quad \Rightarrow \quad -17 = 16 + q \quad \Rightarrow \quad q = -33.$$

Therefore:

$$x^2 + 8x - 17 \equiv (x + 4)^2 - 33.$$

EXAMPLE 4

The polynomial g is defined by $g(x) = (2x - 1)(x^2 + 1) + (5x - 1)(x - 2)$.

Prove that $g(x) \equiv px^3 + qx^2 + rx + s$ where p, q, r and s are constant numbers whose values should be stated.

EXAMPLE 4 (continued)

SOLUTION

Multiplying out the pairs of brackets separately gives

$$(2x-1)(x^2+1) = 2x^3 + 2x - x^2 - 1 = 2x^3 - x^2 + 2x - 1$$

and

$$(5x-1)(x-2) = 5x^2 - 10x - x + 2 = 5x^2 - 11x + 2.$$

Combining these results gives

$$\begin{aligned} g(x) &= (2x-1)(x^2+1) + (5x-1)(x-2) \\ &= 2x^3 - x^2 + 2x - 1 + 5x^2 - 11x + 2 \\ &= 2x^3 + 4x^2 - 9x + 1 \end{aligned}$$

so you have $g(x) = px^3 + qx^2 + rx + s$ where $p = 2$, $q = 4$, $r = -9$, $t = 1$.

EXAMPLE 5

a) Expand and simplify $(3x-1)(x+2)$.

b) The polynomial f is defined by $f(x) = (3x-1)(x+2)$.
Values of $f(x)$ are to be calculated using a computer that takes T seconds to perform an addition or subtraction of two numbers and $2T$ seconds to perform a multiplication of two numbers.

- **i)** Show that this computer will take $6T$ seconds to calculate the value of $f(x)$ using the initial definition $f(x) = (3x-1)(x+2)$.
- **ii)** How long will this computer take to evaluate $f(x)$ if the expanded form of $f(x)$ is used?

SOLUTION

a) $(3x-1)(x+2) = 3x^2 + 6x - x - 2 = 3x^2 + 5x - 2$

b) Evaluating $(3x-1)$ requires a multiplication and a subtraction so takes $3T$ seconds;
Evaluating $(x+2)$ requires an addition so takes T seconds;
Multiplying together $(3x-1)$ and $(x+2)$ requires another $2T$ seconds;
The total time requirement is therefore $6T$.

c) To evaluate $3x^2$ requires a multiplication of x by x followed by a multiplication of the answer by 3 and will therefore take $4T$ seconds.
Evaluation of $5x$ requires a multiplication which will take $2T$ seconds.
Evaluation of $3x^2 + 5x - 2$ now requires the addition of $3x^2$ and $5x$ followed by the subtraction of 2 from the answer and this will take a further $2T$ seconds.
In total, the evaluation of $3x^2 + 5x - 2$ takes $8T$ seconds.

EXERCISE 3

1 Find the expanded form of each of the following polynomials. Simplify your answers as much as possible:

i) $3p(2p+7)$ **ii)** $5p(3p-7)$
iii) $5x^2(x^3+6)$ **iv)** $2y(y-7) + 3y(2y+5)$
v) $(3x-2)(5x+4)$ **vi)** $(p^2-2)(2p+4)$
vii) $(p-4)(p+9)$ **viii)** $(y-9)(y-8)$
ix) $(2t+5)(5t+7)$ **x)** $(2t-9)(2t+9)$
xi) $(3t-1)^2$ **xii)** $(x+5)(x-3) + (x+9)(x-11)$
xiii) $(x-3)^2 + (x+3)^2$ **xiv)** $(x+2)(3x^2+2x-5)$

2 The polynomial g is defined by $g(x) = 5x^2 + 3x - 2$:

a) evaluate $g(-2)$

b) find a simplified expression for $g(x^4)$. What is the degree of $g(x^4)$?

c) Prove that $g(x^2 + 3) = 5x^4 + 33x^2 + 52$

d) If $a > 0$ and $g(ax + b) = 20x^2 - 94x + 108$, find the values of the constants a and b

3 Prove that the following identities are correct:

a) $(5x + 3)^2 + (5x - 3)^2 \equiv 50x^2 + 18$

b) $(x + 5)^2 - (x - 5)^2 \equiv 20x$

c) $(2x + 5)^3 \equiv 8x^3 + 60x^2 + 150x + 125$

4 The polynomial h is defined by $h(x) = x^2(2x + 3)(x - 5)$:

a) expand and simplify the expression $x^2(2x + 3)(x - 5)$

Values of $h(x)$ are to be calculated using a computer that takes T seconds to perform an addition or subtraction of two numbers and $3T$ seconds to perform a multiplication of two numbers.

b) Show that this computer will take $14T$ to evaluate $h(x)$ from the definition $h(x) = x^2(2x + 3)(x - 5)$.

c) How long will this computer take to evaluate $h(x)$ if the expanded form of $f(x)$ is used?

Factorisation of Simple Polynomials

You saw in Example 5 of the previous section that the factorised form of a polynomial may give a more efficient way of calculating values of the polynomial. Moreover, factorisation is often a key step in the solution of equations involving polynomials.

Common Factors

When attempting to factorise an expression you should always start by checking to see whether there are any common factors.

Consider the expression $6y^4 + 4y^2$.

You know that

$$6y^4 = 2y^2 \times 3y^2 \quad \text{and} \quad 4y^2 = 2y^2 \times 2$$

so $2y^2$ is a **common factor** of each term of the expression and you can write

$$6y^4 + 4y^2 = 2y^2(3y^2 + 2).$$

You could also write this result as

$$6y^4 + 4y^2 \equiv 2y^2(3y^2 + 2)$$

if you wished to stress the fact that the result is true for **all values** of y.

Similarly

$$4x^3 + 12x^2 - 8x = 4x(x^2 + 3x - 2).$$

Remember always to find the largest possible common factor.
In this example, you could have written

$$4x^3 + 12x^2 - 8x = 4(x^3 + 3x^2 - 2x)$$

or

$$4x^3 + 12x^2 - 8x = 2x(2x^2 + 6x - 4)$$

and both these factorisations are correct but neither is a full factorisation. Always check your attempt at factorising to make sure that there are no more common factors.

EXERCISE 4

By looking for common factors, factorise the following expressions:

1. $x^2 + 9x$
2. $3x^2 - 6x$
3. $p^2 - 5p$
4. $6p^2 - 18p$
5. $2x^5 + 7x^3$
6. $6x^4 + 12x^2 + 24$
7. $9x^5 - 6x^4 + 21x^3$
8. $8y^5 - 12y^6$
9. $4s^2 + 8s^2$
10. $12t^3 - 18t^2$

Factorisation of Quadratic Expressions of the Form $x^2 + bx + c$

Most polynomials will not have a common factor but they may still be factorisable if two, or more, sets of brackets are used.

Consider the expression $x^2 + 5x - 36$.

The expression $x^2 + 5x - 36$ does not have any common factors.

However, it might have a factorisation involving two brackets of the form

$$(x+p)(x+q)$$

An x is needed in each bracket to ensure that you get an x^2 term when the brackets are multiplied out.

where p and q are numbers.

You want

$$x^2 + 5x - 36 \equiv (x+p)(x+q)$$

but

$$(x+p)(x+q) \equiv x^2 + xq + px + pq \equiv x^2 + (p+q)x + pq$$

so you require

$$x^2 + 5x - 36 \equiv x^2 + (p+q)x + pq.$$

If this is to be true the numbers p and q need to satisfy

$$p + q = 5 \quad \text{and} \quad pq = -36.$$

Since pq is negative, one of the numbers must be positive and the other negative.

You may be able to write down two numbers that satisfy these equations. If not, systematically search through pairs of integers that multiply to -36 until you find the required values:

$pq = -36$			
p	q	$p+q$	$p+q=5$?
1	−36	−35	
−1	36	35	
2	−18	−16	
−2	18	16	
3	−12	−9	
−3	12	9	
4	−9	−5	
−4	9	5	✓

Either way, inspired thought or a systematic search through the factors of −36 leads to the conclusion that the numbers −4 and 9 are the values of p and q.

The factorisation is therefore

$$x^2 + 5x - 36 = (x - 4)(x + 9).$$

EXAMPLE 6

Factorise $x^2 - 15x + 36$.

SOLUTION

$x^2 - 15x + 36$ has no common factor.
Try a two bracket factorisation:

$$x^2 - 15x + 36 \equiv (x + p)(x + q) \equiv x^2 + (p + q)x + pq.$$

If this is to be true you need the numbers p and q to satisfy

$$p + q = -15 \quad \text{and} \quad pq = 36.$$

In this case the two numbers must both be negative since their sum is negative but their product is positive. Informal consideration or a systematic search through the negative integer factors of 36 should lead to the conclusion that the numbers −3 and −12 are the values of p and q.

The factorisation is therefore $x^2 - 15x + 36 = (x - 3)(x - 12)$.

EXAMPLE 7

Factorise $x^2 - 49$.

SOLUTION

Again, $x^2 - 49$ has no common factor.
Trying a two bracket factorisation, you want

$$x^2 - 49 \equiv (x + p)(x + q) \equiv x^2 + (p + q)x + pq.$$

If this is to be true you need the numbers p and q to satisfy

$$p + q = 0 \quad \text{and} \quad pq = -49.$$

It should be clear that the numbers −7 and 7 are the values needed for p and q.

The factorisation is therefore

$$x^2 - 49 = (x - 7)(x + 7).$$

> This is a special case of the **difference of two squares** result:
>
> $$x^2 - a^2 = (x - a)(x + a)$$
>
> Look out for examples like this in the future.

EXAMPLE 8

Factorise $x^2 + 3x + 12$.

SOLUTION

$x^2 + 3x + 12$ has no common factors.

Looking for a two bracket factorisation, you need

$$x^2 + 3x + 12 \equiv (x + p)(x + q) \equiv x^2 + (p + q)x + pq$$

EXAMPLE 8 (continued)

so you are looking for two numbers p and q satisfying

$$p + q = 3 \quad \text{and} \quad pq = 12.$$

No numbers can be found to satisfy these conditions so the expression $x^2 + 3x + 12$ cannot be factorised.

EXERCISE 5

Factorise the following expressions:

1. $x^2 + 11x + 28$
2. $x^2 + 12x + 32$
3. $x^2 + 7x + 12$
4. $x^2 + 20x + 75$
5. $x^2 + 3x - 28$
6. $x^2 - 2x - 15$
7. $x^2 + 8x - 20$
8. $x^2 - 7x - 30$
9. $x^2 - 11x + 30$
10. $x^2 - 8x + 12$
11. $x^2 - 9x + 14$
12. $x^2 + 11x - 60$
13. $x^2 + 8x + 16$
14. $x^2 - 64$
15. $x^2 + x - 30$
16. $x^2 + x - 42$
17. $x^2 - 2x - 48$
18. $x^2 + 3x - 70$
19. $x^2 - 25$
20. $x^2 + 12x + 35$
21. $x^2 + 20x + 100$
22. $x^2 - 12x + 36$
23. $x^2 + 10x - 39$
24. $x^2 - 81$
25. $x^2 - 18x + 80$
26. $x^2 - 121$
27. $x^2 + 10x - 56$

Two Step Factorisations

Sometimes to fully factorise an expression it is necessary to first take out a common factor and then factorise the expression inside the brackets.

EXAMPLE 9

Factorise fully the expression $4y^3 - 16y^2 + 12y$.

SOLUTION

The expression has a common factor of $4y$ so

$$4y^3 - 16y^2 + 12y = 4y(y^2 - 4y + 3).$$

The expression in the brackets can be further factorised into two brackets:

$$4y^3 - 16y^2 + 12y = 4y(y^2 - 4y + 3) = 4y(y - 1)(y - 3).$$

EXAMPLE 10

Factorise fully the expression $6p^3 - 24p$.

SOLUTION

$6p$ is a common factor so

$$6p^3 - 24p = 6p(p^2 - 4).$$

The difference of two squares result can be used to factorise the bracket:

$$6p^3 - 24p = 6p(p^2 - 4) = 6p(p - 2)(p + 2).$$

EXERCISE 6

Factorise fully the following expressions:

1. $3x^3 + 15x^2 - 18x$
2. $4x^3 - 36x$
3. $p^4 - 7p^3 + 10p^2$
4. $5p^3 - 180p$
5. $4y^5 - 16y^3$
6. $6x^3 - 72x^2 + 120x$
7. $8p^2 - 16p - 120$
8. $7z^4 - 28z^2$
9. $2y^3 + 18y^2 + 40y$
10. $5t^2 + 15t - 200$
11. $5t^7 + 30t^6 - 35t^5$
12. $9x^4 - 81x^2$

Factorisation of Quadratic Expressions of the Form $ax^2 + bx + c$

EXAMPLE 11

Factorise the expression $3x^2 + 17x + 10$.

SOLUTION

The expression has no common factors. It may have a factorisation involving two brackets of the form $(3x + p)(x + q)$ where p and q are numbers.

To obtain a $3x^2$ term when the brackets are multiplied out you certainly need $3x$ in one bracket and x in the other.

You want

$$3x^2 + 17x + 10 \equiv (3x + p)(x + q) \equiv 3x^2 + 3qx + px + pq \equiv 3x^2 + (p + 3q)x + pq$$

so you need numbers p and q such that

$$p + 3q = 17 \quad \text{and} \quad pq = 10.$$

Since pq and $p + 3q$ are both positive, you know that p and q must be positive.

You may be able to write down numbers p and q through informal consideration of pairs of numbers whose product is 10. Alternatively, you may need to work systematically through the factors of 10 until you find two numbers that satisfy both conditions:

$pq = 10$			
p	q	$p + 3q$	$p + 3q = 17$?
1	10	31	
10	1	13	
2	5	17	✓

The factorisation is therefore $3x^2 + 17x + 10 = (3x + 2)(x + 5)$.

EXAMPLE 12

Factorise the expression $5x^2 + 7x - 6$.

SOLUTION

The expression has no common factors.
Try a two bracket factorisation:

$$5x^2 + 7x - 6 \equiv (5x + p)(x + q) \equiv 5x^2 + 5qx + px + pq \equiv 5x^2 + (p + 5q)x + pq$$

so you need numbers p and q such that

$$p + 5q = 7 \quad \text{and} \quad pq = -6.$$

EXAMPLE 12 (continued)

In this case one of the two numbers must be positive and the other must be negative since their product is negative.
Informal consideration or a systematic search through pairs of integers that multiply to −6 should lead to the conclusion that −3 and 2 are the numbers required for p and q.
The factorisation is therefore $5x^2 + 7x - 6 = (5x - 3)(x + 2)$.

EXAMPLE 13

Factorise the expression $4x^2 - 16x + 7$.

SOLUTION

The expression has no common factors so you need to look for a two bracket factorisation.
To get $4x^2$ you could have $4x$ in one bracket and x in the other OR you could have $2x$ in each bracket.

First investigate what happens if you have $4x$ and x. You want

$$4x^2 - 16x + 7 \equiv (4x + p)(x + q) \equiv 4x^2 + 4qx + px + pq \equiv 4x^2 + (p + 4q)x + pq$$

so you need numbers p and q such that

$$p + 4q = -16 \quad \text{and} \quad pq = 7.$$

There are no integers that satisfy both these equations. The fact that $pq = 7$ means there are only four possibilities $p = 7,\ q = 1$ or $p = 1,\ q = 7$ or $p = -7,\ q = -1$ or $p = -1,\ q = -7$ and none of these give $p + 4q = -16$.

Consider instead a factorisation of the form $(2x \quad)(2x \quad)$. You want

$$4x^2 - 16x + 7 \equiv (2x + p)(2x + q) \equiv 4x^2 + 2qx + 2px + pq \equiv 4x^2 + (2p + 2q)x + pq$$

so you need numbers p and q such that

$$2p + 2q = -16 \quad \text{and} \quad pq = 7.$$

Again, the fact that $pq = 7$ means there are only four possibilities $p = 7,\ q = 1$ or $p = 1,\ q = 7$ or $p = -7,\ q = -1$ or $p = -1,\ q = -7$.
If $p = -1,\ q = -7$ then $2p + 2q$ is −16 so the factorisation is

$$4x^2 - 16x + 7 = (2x - 1)(2x - 7)$$

EXAMPLE 14

Factorise the expression $4y^2 + 8y - 32$.

SOLUTION

The expression has a common factor of 4 so

$$4y^2 + 8y - 32 = 4(y^2 + 2y - 8)$$

and the expression $y^2 + 2y - 8$ is easily factorised as $(y + 4)(y - 2)$ so

$$4y^2 + 8y - 32 = 4(y^2 + 2y - 8) = 4(y + 4)(y - 2).$$

EXAMPLE 15

Factorise the expression $4p^2 - 10p$.

SOL The expression has common factor $2p$ so $4p^2 - 10p = 2p(2p - 5)$.

Examples 14 and 15 emphasise the importance of looking for a common factor BEFORE looking for two bracket factorisations.

EXERCISE 7

Factorise the following expressions:

1. $2x^2 + 5x + 2$
2. $2x^2 + 7x + 3$
3. $3t^2 + 5t + 2$
4. $2y^2 - 7y - 15$
5. $2x^2 + 8x + 6$
6. $5x^2 - x - 4$
7. $7x^2 - 31x + 12$
8. $3p^2 - 15p - 42$
9. $5x^2 - 7x - 6$
10. $11x^2 - 19x - 6$
11. $6y^2 - 11y + 3$
12. $4x^2 + 7x - 2$
13. $3v^2 - 7v + 2$
14. $10s^2 + 19s - 15$
15. $6w^2 + w - 35$
16. $6x^2 - 18x + 12$
17. $4c^2 + 2c - 30$
18. $6u^2 + 9u + 3$
19. $21x^2 - 17x + 2$
20. $12x^2 - 29x + 15$
21. $8x^2 - 16x$
22. $6x^2 + 9x$
23. $5x^2 - 12x - 9$
24. $6a^2 + 17a + 12$
25. $4x^2 - 24x + 32$
26. $16x^2 - 18x$
27. $8x^2 - 18x + 10$

Further Factorisations

The methods introduced in this chapter can sometimes be applied to polynomials of degree 3 or more. Some examples will simply require the extraction of a common factor as the first stage; other examples will require some initial work before your established techniques can be used.

EXAMPLE 16

Factorise the expression $2p^5 + 10p^4 - 12p^3$.

SOL This expression has a common factor of $2p^3$ so

$$2p^5 + 10p^4 - 12p^3 = 2p^3(p^2 + 5p - 6) = 2p^3(p + 6)(p - 1).$$

EXAMPLE 17

Factorise the expression $x^4 - 3x^2 - 4$.

SOLUTION This expression has no common factors but if you put $u = x^2$ you can see that the expression is really just a quadratic expression in disguise!

$$x^4 - 3x^2 - 4 = u^2 - 3u - 4 = (u + 1)(u - 4)$$

$$\Rightarrow \quad x^4 - 3x^2 - 4 = (x^2 + 1)(x^2 - 4)$$

$$= (x^2 + 1)(x + 2)(x - 2).$$

Difference of two squares: $(x^2 - 4) = (x + 2)(x - 2)$.

Polynomials where the term of highest degree has a negative coefficient can be difficult to factorise directly: it is usually a good idea to start the factorisation by extracting a common factor of -1. For example

$$-x^4 + 3x^2 + 4 = -(x^4 - 3x^2 - 4)$$
$$= -(x^2 + 1)(x + 2)(x - 2).$$

Using the factorisation of example 17.

EXAMPLE 18

Factorise $18x - 12x^2 - 6x^3$.

SOLUTION

$$18x - 12x^2 - 6x^3 = -6x^3 - 12x^2 + 18x$$
$$= -(6x^3 + 12x^2 - 18x)$$
$$= -6x(x^2 + 2x - 3)$$
$$= -6x(x + 3)(x - 1).$$

Rearrange the polynomial so terms are in order with the highest degree terms first.

Take out a factor of -1 to make the first term positive.

Common factor of $6x$.

EXERCISE 8

Factorise fully the expressions in questions 1–8.

1 $x^3 + 7x^2 + 10x$

2 $x^4 - 13x^2 + 36$

3 $2x^3 + 9x^2 - 11x$

4 $27 + 6y^2 - y^4$

5 $3x^4 - 5x^2 + 2$

6 $-10x^5 + 55x^4 + 105x^3$

7 $16 - 6x - x^2$

8 $36 + 5x^2 - x^4$

9 By first letting $u = x^4$ in each case, obtain factorisations for
a) $x^8 + 9x^4 + 20$ **b)** $3x^8 + 11x^8 + 10$ **c)** $x^8 - 14x^4 - 32$ **d)** $x^8 - 1$

10 **a)** Express $x^6 + 9x^3 + 8$ as the product of two cubic polynomials.
b) Express $2x^{10} + 5x^5 - 3$ as the product of two polynomials each of degree 5.
c) Express $6x^{12} + 27x^8 + 21x^4$ as the product of three polynomials each of degree 4.

11 **a)** Factorise $x^6 - 4x^3 - 96$ as the product of two cubic polynomials.
b) Hence use the fact that $8^3 = 512$ to find the value of $8^6 - 4 \times 8^3 - 96$.

12 **a)** Factorise $y^8 - 2y^4 - 24$ as the product of two polynomials of degree 4.
b) Hence find the value of $4^8 - 2 \times 4^4 - 24$.

Having studied this chapter you should know how

- to use function notation
- to interpret and use the $\equiv$ symbol
- to add, subtract and multiply polynomials
- to factorise polynomials which have a common factor

- to factorise quadratic polynomials such as $x^2 + 7x - 30$ and $4x^2 + 11x - 15$ using two sets of brackets
- to factorise fully polynomials such as $2p^3 + 7p^2 - 22p$ by first extracting a common factor and then factorising the expression inside the brackets
- to apply these factorisation methods to expressions that can be regarded as disguised quadratics

REVISION EXERCISE

1 Factorise $x^3 - 16x$ completely

(OCR Nov 1995 P1)

2 Expand and simplify $(x + 1)(x^2 - 2x + 2)$

(OCR Mar 1999 P1)

3 If $f(x) = x^3 + 2x^2 - 5$ and $g(x) = 3x^3 - 4x^2 + 7$
a) find the value of $f(2)$ **b)** simplify $f(x) + g(x)$
c) simplify $3f(x) - g(x)$ **d)** simplify $f(x)g(x)$

4 If $h(p) = 4p(p - 2) + 3p^2 - 5p$
a) write $h(p)$ in its expanded form **b)** write $h(p)$ in its factorised form

5 Expand and simplify
a) $(3p - 2)(2p + 5)$ **b)** $(3p - 2)^2$ **c)** $(3u - 5)(3u + 5)$
d) $2(x + 6)^2$ **e)** $(x + 5)^2 - (x - 5)^2$

6 Factorise
a) $6p^2 - 9pq$ **b)** $p^2 + 7p - 18$ **c)** $y^2 - 64$
d) $y^2 - 11y + 30$ **e)** $5t^2 + 25t - 420$ **f)** $x^4 + 5x^2 - 36$

7 Factorise
a) $3p^2 + 7p - 26$ **b)** $5p^2 + 11p + 2$ **c)** $5y^3 + 10y^2 - 15y$

8 Expand and simplify $(2y - 1)(y^2 + y - 2)$

9 a) Factorise $x^2 + 11x + 24$
b) Hence, without using a calculator, determine the value of $992^2 + 11 \times 992 + 24$

10 The polynomial f is defined by $f(x) = 2x^3 - 3$. Find expressions, not involving brackets, for
a) $6f(x)$ **b)** $(5x - 2)f(x)$ **c)** $(f(x))^2$

11 Factorise fully
a) $6p^3 - 24p$ **b)** $36p^2 - 81p^4$ **c)** $2y^4 - 5y^2 - 12$
d) $t^6 + 10t^3 - 24$ **e)** $64 - 12x^2 - x^4$

12 Prove the identities
a) $3(x - 2)^2 + 13 \equiv 3x^2 - 12x + 25$ **b)** $(2x - 3)^3 \equiv 8x^2 - 36x^2 + 54x - 27$

13 If $2x^2 + 8x - 10 \equiv a(x + b)^2 + c$, find the values of the constants a, b and c.

14 If $9x^2 - 24x + 30 \equiv (px + q)^2 + r$, find the values of the constants p, q and r, given that $p > 0$.

3 Quadratic Equations

The purpose of this chapter is to enable you to

- use the method of factorisation to solve quadratic equations
- use the method of completing the square to solve quadratic equations
- solve quadratic equations by using the quadratic equation formula
- use quadratic equations to solve problems

An equation in an unknown x is a **quadratic equation** if the equation contains an x^2 term together, possibly, with a term involving x and a numerical term.

The following are all examples of quadratic equations:

$x^2 = 9$ $\quad$ $5x^2 = 80$ $\quad$ $x^2 + 7 = 32$

$x^2 + 5x - 6 = 0$ $\quad$ $x^2 = 12 - x$ $\quad$ $1 - 3x^2 = 5x$

Solving Simple Quadratic Equations

The easiest quadratic equations to solve are the equations with x^2 terms but with no x terms in them.

EXAMPLE 1

Solve the equation $4x^2 + 7 = 43$.

SOLUTION

$4x^2 + 7 = 43$

$[-7] \quad \Rightarrow \quad 4x^2 = 36$

$[\div 4] \quad \Rightarrow \quad x^2 = 9$

$\Rightarrow \quad x = 3$ or -3.

EXERCISE 1

Solve the equations

1. $x^2 = 64$
2. $3p^2 - 15 = 33$
3. $2y^2 + 23 = 121$
4. $t^2 - 8 = 73$
5. $\dfrac{u^2}{6} = 1.5$
6. $\dfrac{3p^2 + 22}{5} = 26$

Solving Quadratic Equations by Factorisation

EXAMPLE 2

Solve the equation $x^2 - 6x + 5 = 0$.

SOLUTION

$$x^2 - 6x + 5 = 0$$

Factorising gives $(x - 1)(x - 5) = 0$

$\Rightarrow \quad x - 1 = 0 \quad \text{or} \quad x - 5 = 0$

$\Rightarrow \quad x = 1 \quad \text{or} \quad x = 5.$

The two quantities $(x - 1)$ and $(x - 5)$ multiply together to give 0. **The only way that two numbers can multiply together to give 0 is if at least one of the numbers is itself 0.**

It can easily be checked that both of these solutions do satisfy the original equation:

when $x = 1$: $x^2 - 6x + 5 = 1^2 - 6 \times 1 + 5 = 1 - 6 + 5 = 0$
when $x = 5$: $x^2 - 6x + 5 = 5^2 - 6 \times 5 + 5 = 25 - 30 + 5 = 0$

A **root** of an equation is a value that satisfies the equation.
The **solution** of an equation is a list of **all** of the roots of the equation.

It can be said that 1 is a root of the equation $x^2 - 6x + 5 = 0$.
It can also be said that the solution of the equation $x^2 - 6x + 5 = 0$ is $x = 1$ or $x = 5$.

EXAMPLE 3

Solve the equation $2x^2 - 8x = 120$.

SOLUTION

$2x^2 - 8x = 120$

$[-120] \quad \Rightarrow \quad 2x^2 - 8x - 120 = 0$

$[\div 2] \quad \Rightarrow \quad x^2 - 4x - 60 = 0$

$\Rightarrow \quad (x + 6)(x - 10) = 0$

$\Rightarrow \quad x + 6 = 0 \quad \text{or} \quad x - 10 = 0$

$\Rightarrow \quad x = -6 \quad \text{or} \quad x = 10.$

It is important to remember that this method of solving a quadratic equation relies on having 0 on one side of the quadratic equation.

There is a common factor of 2 in all the terms so each side of the equation can be divided by 2.

Notice that to factorise $2x^2 - 8x - 120$ you would write

$$2x^2 - 8x - 120 = 2(x^2 - 4x - 60) = 2(x + 6)(x - 10)$$

and it would certainly be wrong to divide by 2 at any stage in this process.

On the other hand, dividing each side of an **equation** by 2 to produce a simpler equation is perfectly legitimate.

EXAMPLE 4

Solve the equation $5p^2 = 8p$.

SOLUTION

$5p^2 = 8p$

$[-8p] \quad \Rightarrow \quad 5p^2 - 8p = 0$

$\Rightarrow \quad p(5p - 8) = 0$

$\Rightarrow \quad p = 0 \quad \text{or} \quad 5p - 8 = 0$

$\Rightarrow \quad p = 0 \quad \text{or} \quad p = 1.6.$

You might be tempted to try to solve the equation

$$5p^2 = 8p$$

by dividing each side by p and obtaining

$$5p = 8$$

which then gives

$$p = 1.6.$$

This "method" loses the solution $p = 0$.
The error is dividing the equation by something that could be 0 and division by 0 is impossible!

A general policy for such equations is to collect all the terms on one side and then proceed by factorising.

Remember: **FACTORISE don't divide!**

EXAMPLE 5

Solve the equation $3x^2 = 8 - 5x$.

SOLUTION

$$\begin{aligned} & & 3x^2 &= 8 - 5x \\ [+5x] & \Rightarrow & 3x^2 + 5x &= 8 \\ [-8] & \Rightarrow & 3x^2 + 5x - 8 &= 0 \\ & \Rightarrow & (3x+8)(x-1) &= 0 \\ & \Rightarrow & 3x + 8 = 0 \quad &\text{or} \quad x - 1 = 0 \\ & \Rightarrow & 3x = -8 \quad &\text{or} \quad x = 1 \\ & \Rightarrow & x = \frac{-8}{3} \quad &\text{or} \quad x = 1. \end{aligned}$$

Again, the first aim must be to collect all the terms on one side of the equation and have a 0 on the other side.

EXAMPLE 6

Solve the equation $x(6x - 5) = 14$.

SOLUTION

$$\begin{aligned} & x(6x-5) = 14 \\ \Rightarrow\ & 6x^2 - 5x = 14 \\ \Rightarrow\ & 6x^2 - 5x - 14 = 0 \\ \Rightarrow\ & (6x+7)(x-2) = 0 \\ \Rightarrow\ & 6x + 7 = 0 \quad \text{or} \quad x - 2 = 0 \\ \Rightarrow\ & x = -\tfrac{7}{6} \quad \text{or} \quad x = 2. \end{aligned}$$

It is quite acceptable to leave an answer as a top heavy fraction.

EXAMPLE 7

If 3 is a root of the equation $x^2 + bx - 21 = 0$ find the value of the constant b and determine the other root of the equation.

SOLUTION

Since 3 is a root of the equation

$$\begin{aligned} & 3^2 + b \times 3 - 21 = 0 \\ \Rightarrow\ & 9 + 3b - 21 = 0 \\ \Rightarrow\ & 3b - 12 = 0 \\ \Rightarrow\ & b = 4. \end{aligned}$$

The equation is therefore

$$\begin{aligned} & x^2 + 4x - 21 = 0 \\ \Rightarrow\ & (x+7)(x-3) = 0 \\ \Rightarrow\ & x + 7 = 0 \quad \text{or} \quad x - 3 = 0 \\ \Rightarrow\ & x = -7 \quad \text{or} \quad x = 3. \end{aligned}$$

The other root of the equation is −7.

EXERCISE 2

1 Solve the equations

a) $x^2 - 5x - 6 = 0$ **b)** $2x^2 - 3x - 2 = 0$ **c)** $x^2 - 9x = 0$

d) $x^2 - x - 12 = 0$ **e)** $6t^2 - 5t + 1 = 0$ **f)** $3p^2 + p - 2 = 0$

2 **a)** Find the roots of the equation $x^2 - 16x + 60 = 0$.
b) Find the roots of the equation $4u^2 + 7u - 30 = 0$.

3 Find the solution of the equations
a) $x^2 - 16x = 80$ **b)** $x^2 = 7x$ **c)** $x^2 + 4x = 77$

4 **a)** Factorise fully the expression $3x^2 + 15x - 42$.
b) Hence find the roots of the equation $3x^2 + 15x - 42 = 0$.

5 **a)** Factorise fully the expression $5x^2 - 45$.
b) Hence, or otherwise, find the roots of the equation $5x^2 - 45 = 0$.

6 Solve the equations
a) $x(x+5) = 3x + 8$ **b)** $x(x-6) = 27$ **c)** $5x(x+7) = 90$
d) $13 - 5y^2 - 8y = 0$ **e)** $2q^2 = 7q - 6$ **f)** $x(2x+5) = 3$

7 Find the solutions of the equations
a) $(2x+1)(x-3) = 22$ **b)** $4r^2 = 11r - 7$ **c)** $5x(x-2) = 2x(x+6)$
d) $7x^2 = 20x + 3$ **e)** $3t(t-2) = t^2 + 8$

8 **a)** If -2 is a root of the equation $ax^2 + 9x + 10 = 0$ find the value of a and determine the second root of the equation.
b) If 5 is a root of the equation $2x^2 - 3x + c = 0$ find the value of c and determine the second root of the equation.

Completing the Square

The Completed Square Format of $x^2 + bx + c$

So far, the only quadratic equations that you have been able to solve are ones where it is possible to factorise the quadratic expression. However, many quadratic equations cannot be solved by factorisation.

For example, consider the equation $x^2 + 8x - 13 = 0$.
It is not possible to factorise the left-hand side of the expression since there are no two integers that multiply to -13 and add to 8.
However, the graph of $y = x^2 + 8x - 13$ makes it clear that the equation does have two roots.

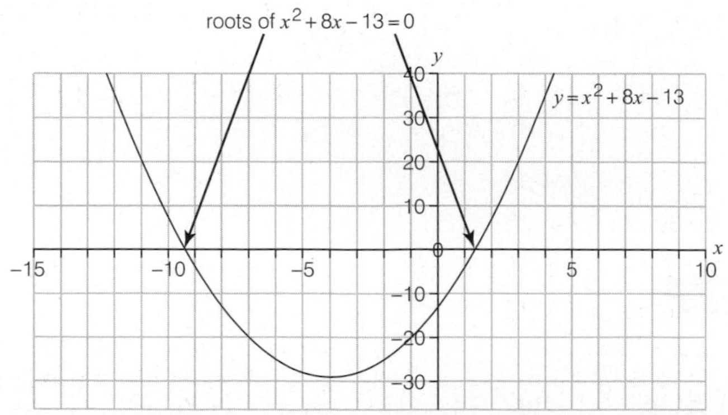

These roots can be found by a process known as **completing the square**.

Consider the expression $x^2 + 8x$.

From the diagram you can see that $x^2 + 8x$ is the area of the large square minus the shaded area:

	x	4
x	x^2	$4x$
4	$4x$	**16**

This suggests the result

$$x^2 + 8x = (x + 4)^2 - 16.$$

This can be checked by expanding the brackets on the right-hand side:

$$\begin{aligned}(x + 4)^2 - 16 &= (x + 4)(x + 4) - 16\\ &= x^2 + 4x + 4x + 16 - 16\\ &= x^2 + 8x\end{aligned}$$

as required.

It can be said that $(x + 4)^2 - 16$ is the **completed square form** of $x^2 + 8x$.

Thus $x^2 + 8x - 13 = 0$
can now be rewritten as $(x + 4)^2 - 16 - 13 = 0$

$\Rightarrow\quad (x + 4)^2 = 29$

$\Rightarrow\quad x + 4 = \sqrt{29}$ or $-\sqrt{29}$

$\Rightarrow\quad x = -4 + \sqrt{29}$ or $-4 - \sqrt{29}$ — These are the **exact** solutions of the equation.

$\Rightarrow\quad x = 1.385$ or $x = -9.385$ (4 s.f.).

The procedure of completing the square gives us a method of finding the solutions of any quadratic equation. It does involve more working than a solution by factorisation so it is always worth first looking to see whether or not the equation will factorise easily.

EXAMPLE 8

Write $x^2 + 13x$ in completed square form and hence solve the equation

$$x^2 + 13x - 27 = 0.$$

SOLUTION

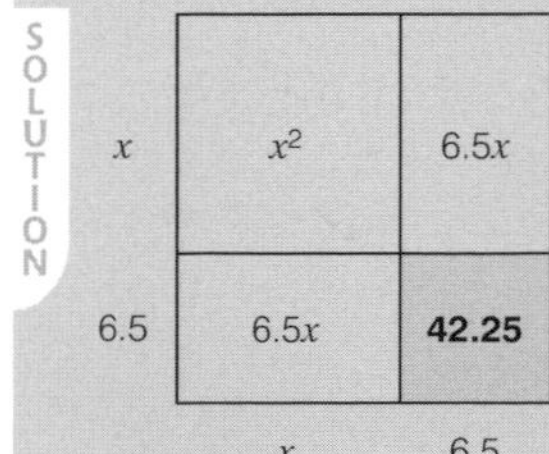

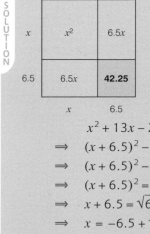

From the diagram it is seen that

$$x^2 + 13x = (x + 6.5)^2 - 42.25$$

and this could be checked by expanding the brackets on the right-hand side.

The completed square form is now used to solve the equation:

$x^2 + 13x - 27 = 0$

$\Rightarrow\quad (x + 6.5)^2 - 42.25 - 27 = 0$

$\Rightarrow\quad (x + 6.5)^2 - 69.25 = 0$

$\Rightarrow\quad (x + 6.5)^2 = 69.25$

$\Rightarrow\quad x + 6.5 = \sqrt{69.25}$ or $-\sqrt{69.25}$

$\Rightarrow\quad x = -6.5 + \sqrt{69.25}$ or $x = -6.5 - \sqrt{69.25}$ — These values are **exact**.

$\Rightarrow\quad x = 1.822$ or $x = -14.82$ (4 s.f.).

EXAMPLE 9

Write $x^2 - 6x$ in completed square form and hence find the exact values of the roots of the equation $x^2 - 6x = 2$.

SOLUTION

	x	-3
x	x^2	$-3x$
-3	$-3x$	**9**

From the diagram it is seen that

$$x^2 - 6x = (x + (-3))^2 - 9 = (x - 3)^2 - 9$$

and this could be checked by expanding the brackets on the right-hand side:

$$x^2 - 6x = 2$$
$$\Rightarrow \quad (x-3)^2 - 9 = 2$$
$$\Rightarrow \quad (x-3)^2 = 11$$
$$\Rightarrow \quad x - 3 = \sqrt{11} \quad \text{or} \quad -\sqrt{11}$$
$$\Rightarrow \quad x = 3 + \sqrt{11} \quad \text{or} \quad x = 3 - \sqrt{11}.$$

The next example suggests an alternative approach to finding the completed square form.

EXAMPLE 10

Find constants p and q so that $x^2 + 16x + 5 \equiv (x+p)^2 + q$ and hence deduce the exact roots of the equation $x^2 + 16x + 5 = 0$.

SOLUTION

You know that $\quad (x+p)^2 + q \equiv (x+p)(x+p) + q \equiv x^2 + 2px + p^2 + r$

so you want $\quad x^2 + 16x + 5 \equiv x^2 + 2px + p^2 + r$.

In the previous chapter you saw that two polynomials are identically equal only if all of the coefficients agree:

Looking at x coefficients: $\quad 16 = 2p \quad \Rightarrow \quad p = 8$

Looking at the constant terms: $\quad 5 = p^2 + r \quad \Rightarrow \quad 5 = 64 + r \quad \Rightarrow \quad r = -59.$

Therefore $\quad x^2 + 16x + 5 \equiv (x+8)^2 - 59$

so the equation $\quad x^2 + 16x + 5 = 0$

can be written as $\quad (x+8)^2 - 59 = 0$

$$\Rightarrow \quad (x+8)^2 = 59$$
$$\Rightarrow \quad x + 8 = \sqrt{59} \quad \text{or} \quad -\sqrt{59}$$
$$\Rightarrow \quad x = -8 + \sqrt{59} \quad \text{or} \quad -8 - \sqrt{59}.$$

EXERCISE 3

1 Write the following expressions in their completed square forms:

a) $x^2 + 10x$ **b)** $x^2 + 4x$ **c)** $x^2 + 14x$ **d)** $x^2 + 5x$ **e)** $x^2 + 7x$

f) $x^2 - 10x$ **g)** $x^2 - 16x$ **h)** $x^2 - 9x$ **i)** $x^2 - 20x$ **j)** $x^2 + 12x$

What is the completed square form for $x^2 + px$?

2 Find the exact solutions of the following quadratic equations by using the method of completing the square:

a) $x^2 + 8x - 15 = 0$ **b)** $x^2 + 18x - 35 = 0$

c) $x^2 - 12x + 9 = 0$ **d)** $x^2 - 7x + 2 = 0$

3 Find constants p and q so that $x^2 - 8x + 5 \equiv (x + p)^2 + q$ and hence deduce the exact roots of the equation

$$x^2 - 8x + 5 = 0$$

4 Find constants p and q so that $x^2 + 4x - 17 \equiv (x + p)^2 + q$ and hence deduce the exact roots of the equation

$$x^2 + 4x - 17 = 0$$

5 Find the solutions of the following quadratic equations by using the method of completing the square. Give your answers correct to four significant figures:

a) $x^2 - 9x = 35$ **b)** $x^2 + 3x = 10$

c) $x^2 + 0.8x - 1.93 = 0$ **d)** $x^2 + 84x = -1500$

Solving the Equation $ax^2 + bx + c = 0$

The method introduced in the previous section can be extended to quadratic equations of the form $ax^2 + bx + c = 0$.

EXAMPLE 11

Solve the equation $2x^2 + 10x - 19 = 0$.

SOLUTION

$$2x^2 + 10x - 19 = 0$$

$$[\div 2] \quad \Rightarrow \quad x^2 + 5x - 9.5 = 0$$

and the process of completing the square gives

$$x^2 + 5x = (x + 2.5)^2 - 6.25$$

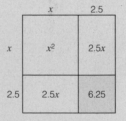

	x	2.5
x	x^2	$2.5x$
2.5	$2.5x$	6.25

so

$$(x + 2.5)^2 - 6.25 - 9.5 = 0$$

$$\Rightarrow \quad (x + 2.5)^2 = 15.75$$

$$\Rightarrow \quad x + 2.5 = \sqrt{15.75} \quad \text{or} \quad -\sqrt{15.75}$$

$$\Rightarrow \quad x = -2.5 + \sqrt{15.75} \quad \text{or} \quad x = -2.5 - \sqrt{15.75}$$

$$\Rightarrow \quad x = 1.469 \quad \text{or} \quad x = -6.469 \qquad (4 \text{ s.f.}).$$

EXERCISE 4

Solve the following quadratic equations, giving your answers correct to four significant figures:

1 $5x^2 + 25x - 37 = 0$ **2** $4x^2 - 24x + 17 = 0$

3 $8x^2 + 25x = 57$ **4** $15x^2 - 36x = 157$

Find the exact solutions of the following quadratic equations:

5 $2x^2 - 8x - 11 = 0$

6 $5x^2 - 12x + 2 = 0$

7 $0.2x^2 + 0.8x - 6 = 0$

8 $\frac{1}{4}x^2 + 7x - 8 = 0$

The Formula for Solving Quadratic Equations

The method of completing the square allows the development of a general formula for solving all quadratic equations.

Consider the quadratic equation

$$ax^2 + bx + c = 0$$

$$[\div a] \quad \Rightarrow \quad x^2 + \frac{b}{a}x + \frac{c}{a} = 0.$$

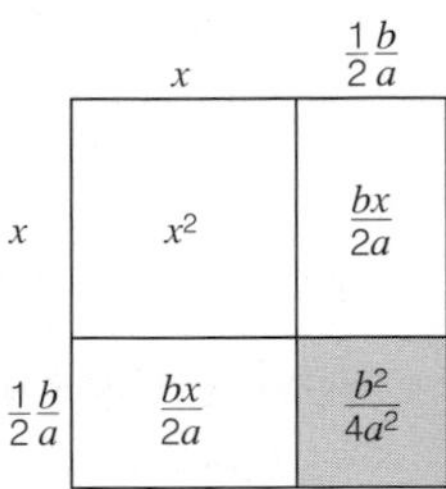

Consider now the expression $x^2 + \frac{b}{a}x$.

Completing the square gives

$$x^2 + \frac{b}{a}x = \left(x + \frac{1}{2}\frac{b}{a}\right)^2 - \frac{b^2}{4a^2}.$$

The equation now becomes

$$\left(x + \frac{1}{2}\frac{b}{a}\right)^2 - \frac{b^2}{4a^2} + \frac{c}{a} = 0$$

$$\Rightarrow \quad \left(x + \frac{1}{2}\frac{b}{a}\right)^2 = \frac{b^2}{4a^2} - \frac{c}{a}$$

Common denominator $= 4a^2$.

$$\Rightarrow \quad \left(x + \frac{1}{2}\frac{b}{a}\right)^2 = \frac{b^2 - 4ac}{4a^2}$$

$$\Rightarrow \quad \left(x + \frac{1}{2}\frac{b}{a}\right) = \sqrt{\frac{b^2 - 4ac}{4a^2}} \quad \text{or} \quad -\sqrt{\frac{b^2 - 4ac}{4a^2}}$$

$$\Rightarrow \quad \left(x + \frac{1}{2}\frac{b}{a}\right) = \frac{\sqrt{b^2 - 4ac}}{2a} \quad \text{or} \quad -\frac{\sqrt{b^2 - 4ac}}{2a}$$

$$\Rightarrow \quad x = -\frac{b}{2a} + \frac{\sqrt{b^2 - 4ac}}{2a} \quad \text{or} \quad -\frac{b}{2a} - \frac{\sqrt{b^2 - 4ac}}{2a}$$

$$\Rightarrow \quad x = \frac{-b + \sqrt{b^2 - 4ac}}{2a} \quad \text{or} \quad \frac{-b - \sqrt{b^2 - 4ac}}{2a}.$$

The two solutions can be summarised in a single formula:

> If
> $$ax^2 + bx + c = 0$$
> then
> $$x = \frac{-b \pm \sqrt{b^2 - 4ac}}{2a}$$

Many graphical calculators will find the solutions of the quadratic equation $ax^2 + bx + c = 0$ by using this formula but **you need to know this formula and be able to use it properly without a calculator to obtain exact solutions**.

EXAMPLE 12

Solve the equation $3x^2 + 5x - 10 = 0$.

SOLUTION

Using $a = 3$, $b = 5$ and $c = -10$ obtains

$$x = \frac{-5 \pm \sqrt{5^2 - 4 \times 3 \times (-10)}}{2 \times 3} = \frac{-5 \pm \sqrt{25 - (-120)}}{6}$$

Take great care with the arithmetic inside the square root!

$$x = \frac{-5 + \sqrt{145}}{6} \quad \text{or} \quad \frac{-5 - \sqrt{145}}{6}$$

These are the **exact** solutions.

$$\Rightarrow \quad x = 1.174 \quad \text{or} \quad -2.840 \qquad (4 \text{ s.f.}).$$

EXAMPLE 13

Solve the equation $5x(x - 3) = 8x - 10$, giving your answers correct to four significant figures.

SOLUTION

The equation must first be rewritten in the standard "$ax^2 + bx + c = 0$" format:

$$5x(x - 3) = 8x - 10$$

$$\Rightarrow \quad 5x^2 - 15x = 8x - 10$$

$$\Rightarrow \quad 5x^2 - 23x + 10 = 0.$$

Using $a = 5$, $b = -23$ and $c = 10$ obtains

$$x = \frac{-(-23) \pm \sqrt{(-23)^2 - 4 \times 5 \times 10}}{2 \times 5} = \frac{23 \pm \sqrt{529 - 200}}{10}$$

$$= \frac{23 \pm \sqrt{329}}{10}$$

$$\Rightarrow \quad x = 4.114 \quad \text{or} \quad 0.4862 \qquad (4 \text{ s.f.}).$$

EXERCISE 5

In questions 1–5, solve the quadratic equations, giving the answers correct to three significant figures. (You could use a graphical calculator to check your answers.)

1 $x^2 + 3x + 1 = 0$

2 $x^2 + x - 1 = 0$

3 $3x^2 - 6x - 2 = 0$

4 $6x^2 = 13x + 12$

5 $-4.9t^2 + 15t = 8$

In questions 6–10, find the exact values of the roots of the quadratic equations, without using a calculator.

6 $x^2 - 2x - 4 = 0$

7 $2x^2 - 4x + 1 = 0$

8 $3x^2 + 8x = 9$

9 $3x(x - 1) = 4$

10 $4x(x - 3) = (2x - 3)(x - 2)$

Using Quadratic Equations to Solve Problems

Quadratic equations arise as the final mathematical formulation of a wide variety of problems. In this section and exercise 6 you will meet some examples which lead, eventually, to the solution of a quadratic equation.

EXAMPLE 14

A rectangular plot of land has length which is 3 m less than twice the width. The area of the plot is 77 m^2. If w is the width of the plot in metres:

a) Prove that w must satisfy the equation $2w^2 - 3w - 77 = 0$.
b) Hence find the dimensions of the plot.

SOLUTION

a) If the width of the plot is w then the length of the plot is $2w - 3$.

$$\begin{aligned}\text{Area} &= \text{width} \times \text{length} \\ &= w(2w-3) \\ &= 2w^2 - 3w.\end{aligned}$$

Width = w | Area = 77 | Length = $2w - 3$

You know the area is 77 so

$$\begin{aligned}& 2w^2 - 3w = 77 \\ \Rightarrow\quad & 2w^2 - 3w - 77 = 0.\end{aligned}$$

b) The value of w can be found by solving this quadratic equation:

$$\begin{aligned}& 2w^2 - 3w - 77 = 0 \\ \Rightarrow\quad & (2w+11)(w-7) = 0 \\ \Rightarrow\quad & 2w + 11 = 0 \quad \text{or} \quad w - 7 = 0 \\ \Rightarrow\quad & w = -\tfrac{11}{2} \quad \text{or} \quad w = 7.\end{aligned}$$

If you can't find the factorisation then you could always solve the equation either by completing the square or by using the quadratic equation formula.

The first solution can be rejected since the width, w, must be positive

$$\Rightarrow \quad w = 7$$

The plot therefore has width 7 m and length 11 m.

EXAMPLE 15

A rectangular room is 7 metres wider than it is high, and it is 5 metres longer than it is wide.
If the height of the room is h metres, find expressions, involving h, for

i) the width of the room;
ii) the length of the room;
iii) the total area of the walls of the room.

If the total area of the walls of the room is 150 m^2, write down an equation that must be satisfied by h and solve the equation.
Calculate the volume of the room.

EXAMPLE 15 (continued)

SOLUTION

i) width = height + 7 = $h + 7$
ii) length = width + 5 = $h + 7 + 5 = h + 12$
iii) A sketch of the room can now be drawn:

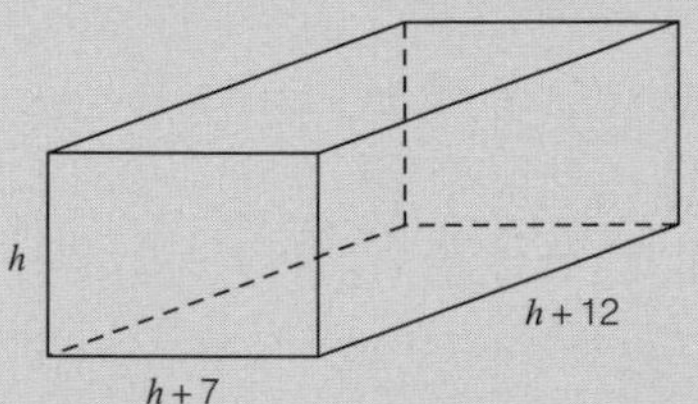

$$\begin{aligned}\text{Wall area} &= 2 \times \text{width} \times \text{height} + 2 \times \text{length} \times \text{height}\\ &= 2(h+7)h + 2(h+12)h\\ &= 2h^2 + 14h + 2h^2 + 24h\\ &= 4h^2 + 38h.\end{aligned}$$

The total wall area is to be 150 m^2 so

$$\begin{aligned}& & 4h^2 + 38h &= \text{wall area} = 150\\ [-150] & \Rightarrow & 4h^2 + 38h - 150 &= 0\\ [\div 2] & \Rightarrow & 2h^2 + 19h - 75 &= 0\\ & \Rightarrow & (2h+25)(h-3) &= 0\\ & \Rightarrow & 2h + 25 = 0 \quad &\text{or} \quad h - 3 = 0\\ & \Rightarrow & h = -12.5 \quad &\text{or} \quad h = 3.\end{aligned}$$

There is a common factor of 2 in each term so you divide each side of the equation by 2.

Since h is the height of the room it must be positive so the "$h = -12.5$" solution can be rejected and you can conclude that $h = 3$.

The room therefore has dimensions:

height = 3 m, width = 10 m, length = 15 m

so

volume = $3 \times 10 \times 15 = 450$ m^3.

EXAMPLE 16

A rectangular lawn measuring 8 m by 12 m is surrounded by a paved path of constant width w metres. The area of the path is 23 m^2. Form a quadratic equation that must be satisfied by w and hence find the width of the path.

SOLUTION

Start by sketching the lawn and the path:

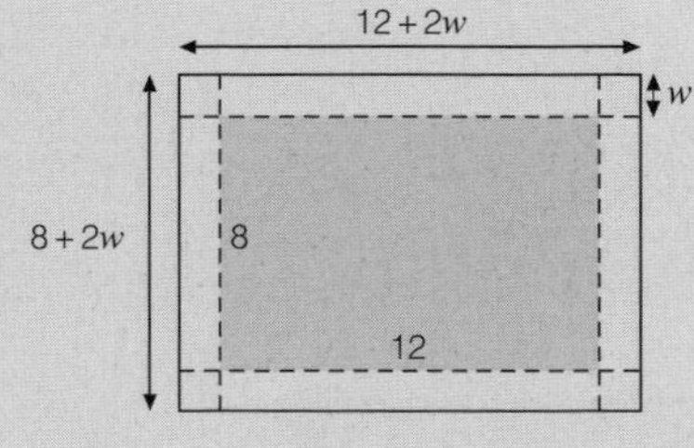

Area of lawn = $8 \times 12 = 96$ m^2.
Area of lawn and paving = $(12 + 2w)(8 + 2w)$.

So

$$\begin{aligned}\text{Area of paving} &= (12+2w)(8+2w) - 96\\ &= 4w^2 + 40w + 96 - 96\\ &= 4w^2 + 40w.\end{aligned}$$

You are told the area of the paving is 23 m^2 so

$$\begin{aligned}& 4w^2 + 40w = 23\\ \Rightarrow\ & 4w^2 + 40w - 23 = 0.\end{aligned}$$

EXAMPLE 16 (continued)

This expression cannot be factorised: the quadratic equation formula gives

$$w = \frac{-40 \pm \sqrt{40^2 - 4 \times 4 \times (-23)}}{2 \times 4} = \frac{-40 \pm \sqrt{1600 - (-368)}}{8}$$

$$\Rightarrow \quad w = \frac{-40 \pm \sqrt{1968}}{8} = 0.5453 \quad \text{or} \quad -10.55 \qquad (4 \text{ s.f.}).$$

The second solution can be rejected since the width, w, must be positive.

$$\Rightarrow \quad w = 0.5453 \qquad (4 \text{ s.f.}).$$

The path has width 0.5453 m, correct to four significant figures.

EXAMPLE 17

The perimeter of a rectangle is 40 cm long and its area is 51 cm^2.
Find the dimensions of the rectangle.

SOLUTION

Let

W = the width of the rectangle
L = the length of the rectangle

then

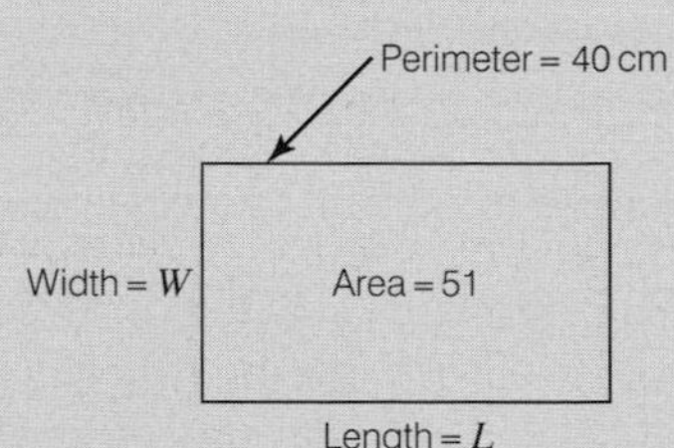

$$40 = \text{perimeter} = 2W + 2L$$
$$\Rightarrow \quad 20 = W + L$$
$$\Rightarrow \quad L = 20 - W$$

and

$$51 = \text{area} = W \times L$$
$$\Rightarrow \quad 51 = W(20 - W)$$
$$\Rightarrow \quad 51 = 20W - W^2$$
$$\Rightarrow \quad W^2 - 20W + 51 = 0$$
$$\Rightarrow \quad (W - 3)(W - 17) = 0$$
$$\Rightarrow \quad W - 3 = 0 \quad \text{or} \quad W - 17 = 0$$
$$\Rightarrow \quad W = 3 \qquad \text{or} \quad W = 17.$$

If $W = 3$ then $L = 17$; if $W = 17$ then $L = 3$.
In both cases you have a rectangle whose sides are 3 cm and 17 cm long.

EXERCISE 6

1 A rectangular field is 6 m longer than it is wide and has an area of 112 m^2.
Suppose that the field is w metres wide.
Show that w must satisfy the equation

$$w^2 + 6w - 112 = 0.$$

Hence find the width and the perimeter of the field.

w 112 m^2

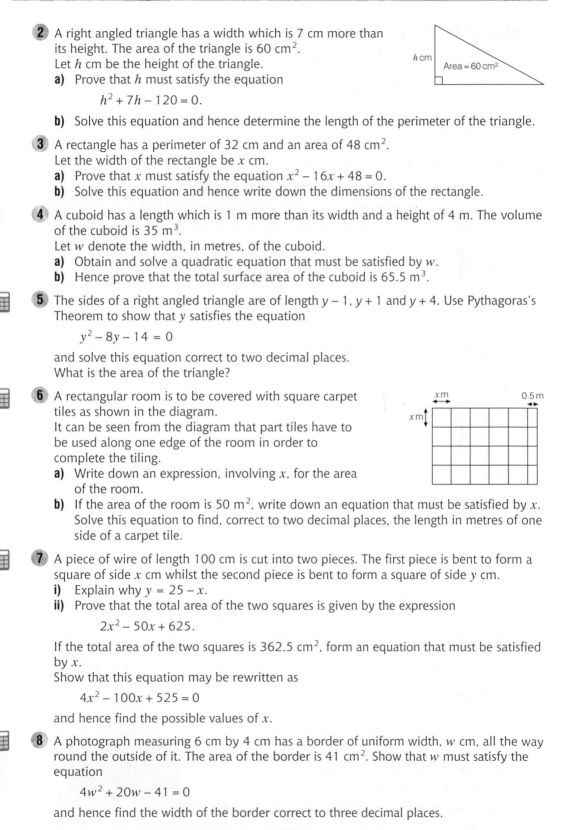

2 A right angled triangle has a width which is 7 cm more than its height. The area of the triangle is 60 cm^2.
Let h cm be the height of the triangle.

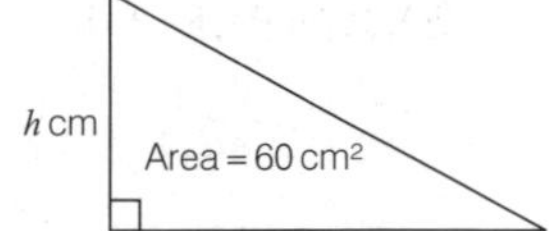

a) Prove that h must satisfy the equation

$$h^2 + 7h - 120 = 0.$$

b) Solve this equation and hence determine the length of the perimeter of the triangle.

3 A rectangle has a perimeter of 32 cm and an area of 48 cm^2.
Let the width of the rectangle be x cm.

a) Prove that x must satisfy the equation $x^2 - 16x + 48 = 0$.

b) Solve this equation and hence write down the dimensions of the rectangle.

4 A cuboid has a length which is 1 m more than its width and a height of 4 m. The volume of the cuboid is 35 m^3.
Let w denote the width, in metres, of the cuboid.

a) Obtain and solve a quadratic equation that must be satisfied by w.

b) Hence prove that the total surface area of the cuboid is 65.5 m^3.

5 The sides of a right angled triangle are of length $y - 1$, $y + 1$ and $y + 4$. Use Pythagoras's Theorem to show that y satisfies the equation

$$y^2 - 8y - 14 = 0$$

and solve this equation correct to two decimal places.
What is the area of the triangle?

6 A rectangular room is to be covered with square carpet tiles as shown in the diagram.
It can be seen from the diagram that part tiles have to be used along one edge of the room in order to complete the tiling.

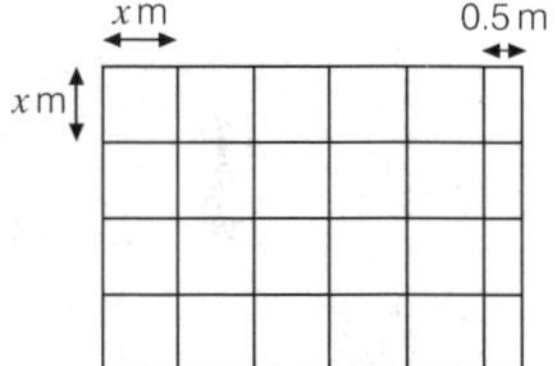

a) Write down an expression, involving x, for the area of the room.

b) If the area of the room is 50 m^2, write down an equation that must be satisfied by x. Solve this equation to find, correct to two decimal places, the length in metres of one side of a carpet tile.

7 A piece of wire of length 100 cm is cut into two pieces. The first piece is bent to form a square of side x cm whilst the second piece is bent to form a square of side y cm.

i) Explain why $y = 25 - x$.

ii) Prove that the total area of the two squares is given by the expression

$$2x^2 - 50x + 625.$$

If the total area of the two squares is 362.5 cm^2, form an equation that must be satisfied by x.
Show that this equation may be rewritten as

$$4x^2 - 100x + 525 = 0$$

and hence find the possible values of x.

8 A photograph measuring 6 cm by 4 cm has a border of uniform width, w cm, all the way round the outside of it. The area of the border is 41 cm^2. Show that w must satisfy the equation

$$4w^2 + 20w - 41 = 0$$

and hence find the width of the border correct to three decimal places.

 **9** A trapezium ABCD has ∠ABC = ∠BCD = 90°. BC = x cm.
AB and DC are parallel and of length x cm and $(x + 7)$ cm, respectively.
If the area of the trapezium is 60 cm^2:
a) Prove that $2x^2 + 7x = 120$.
b) Hence find the length, correct to three significant figures, of BC.

 **10** A garden has two circular ponds. The smaller pond has radius r metres and the larger pond has radius $(r + 2)$ metres. If the total area of the two ponds is 100 m^2, find the radius of each of the ponds, giving your answers to three significant figures.

Having studied this chapter you should know how

- to solve a quadratic equation using the method of factorisation
- to rewrite a quadratic expression of the form $x^2 + bx + c$ in a completed square form $(x + p)^2 + q$
- to use the method of completing the square to solve quadratic equations
- to use the formula $x = \dfrac{-b \pm \sqrt{b^2 - 4ac}}{2a}$ to solve the equation $ax^2 + bx + c = 0$
- to express problems as quadratic equations and hence solve them

REVISION EXERCISE

1 Solve the quadratic equations
a) $5x^2 - 7 = 73$ **b)** $(2x - 1)(x + 3) = 12x + 1$
c) $(x + 3)^2 = (2x + 3)(x + 1)$ **d)** $(3x - 2)^2 = (x + 6)^2$

2 The polynomial f(x) is defined by $f(x) = x^2 + 3x - 88$. Find the solutions of the equation $f(x) = 0$.

3 **a)** Factorise $2x^2 + 5x - 7$.
b) Hence solve the equation $2x^2 + 5x - 7 = 0$.

4 If 7 is a root of the equation $3x^2 - 20x + c = 0$ find the value of the constant c and determine the other root of the equation.

5 The diagram shows a photograph measuring 15 cm by 10 cm surrounded by a frame of width w cm. The area of the frame is 84 cm^2.
Prove that w must satisfy the equation

$$2w^2 + 25w = 42.$$

Solve this equation and hence find the value of w.

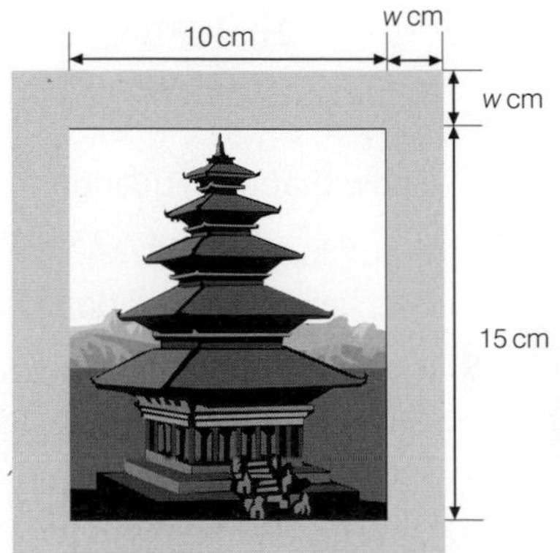

6 Find the solutions, correct to three decimal places, of

a) $2x^2 - 7x - 12 = 0$

b) $(3x + 2)(x - 1) = 7x + 23$

7 Find constants p and q so that $x^2 + 10x - 7 = (x + p)^2 + q$.
Hence find the exact values of the roots of the equation $x^2 + 10x - 7 = 0$.

8 The diagram shows a trapezium ABCD of area $50\ \text{cm}^2$ and height h cm.

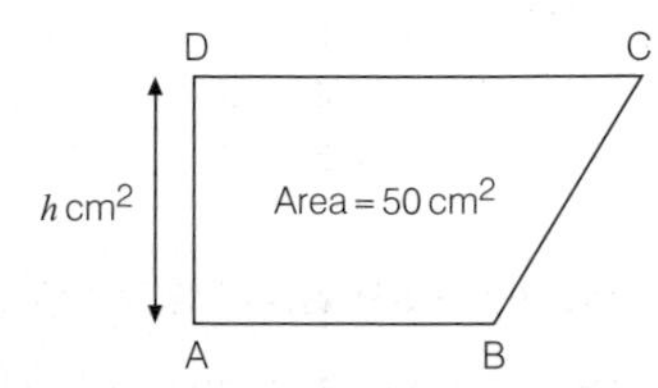

AB = h; CD = AB + 3.

Obtain a quadratic equation that must be satisfied by h and hence find the height of the trapezium.

9 The diagram shows a lawn of area $650\ \text{m}^2$ in an ornamental garden which consists of two semi-circular regions at each end and a rectangle in the middle.

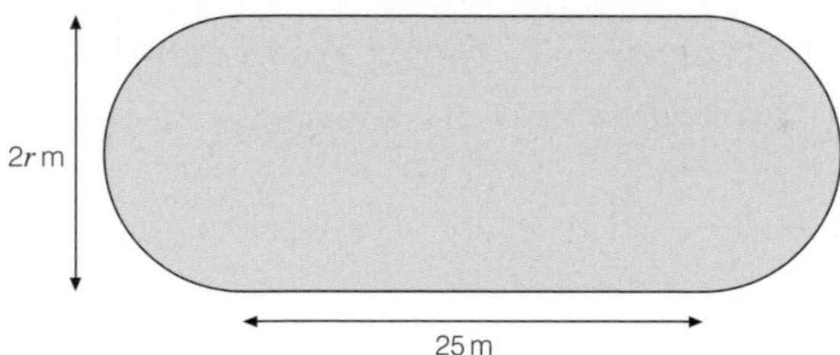

If the radius of the semi-circles is r and the length of the rectangle is 25 m, calculate the value of r and the length of the perimeter of the lawn, giving your answer correct to three significant figures.

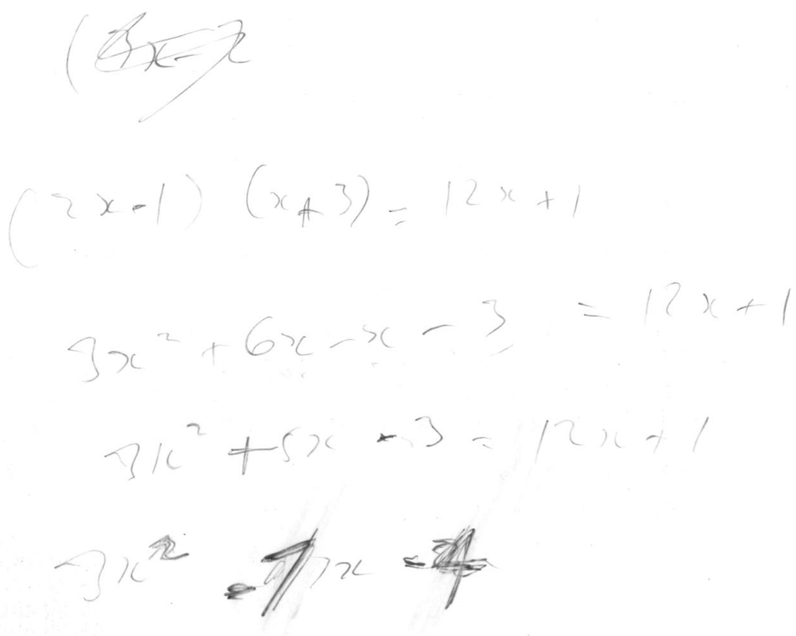

4 Co-ordinate Geometry: Straight Lines

The purpose of this chapter is to enable you to

- use vectors to describe movements on the (x, y) plane
- find the distance between two points and the midpoint of the line segment joining two points
- find the gradient of the line joining two points
- understand and use the equation of a straight line
- find the condition for two lines to be parallel and the condition for two lines to be perpendicular
- find the point of intersection of two lines

Introductory Ideas

Vector Notation

If A is the point (2, 5) and B is the point (5, 3) then $\overrightarrow{\mathbf{AB}}$ is the directed line segment or **vector** that joins A to B.

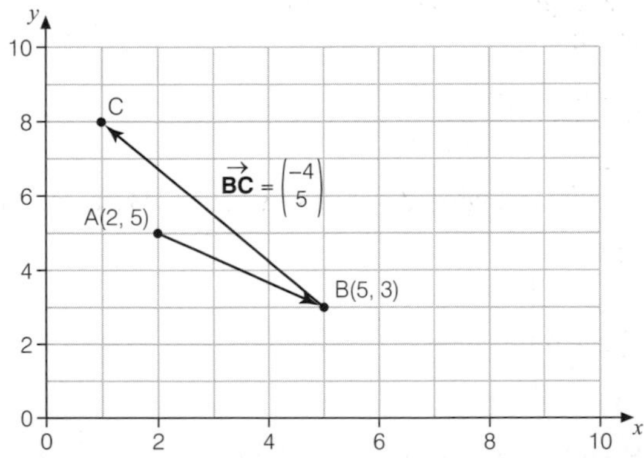

In moving from A to B it is necessary to move 3 in the x direction and −2 in the y direction.

This is written $\overrightarrow{\mathbf{AB}} = \begin{pmatrix} 3 \\ -2 \end{pmatrix}$

If $\overrightarrow{\mathbf{BC}} = \begin{pmatrix} -4 \\ 5 \end{pmatrix}$ then, in moving from B(5, 3) to C the x co-ordinate changes by −4, so the x co-ordinate of C must be 1, and the y co-ordinate changes by 5, so the y co-ordinate of C must be 8.
C is therefore the point (1, 8).

The Distance Between Two Points and the Midpoint of the Line Segment Joining Two Points

If A is the point (5, 1) and B is the point (11, 9) then

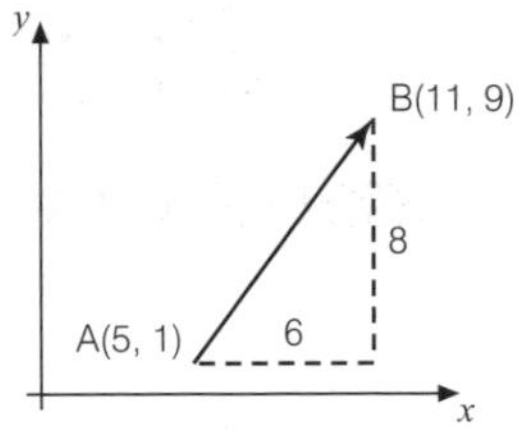

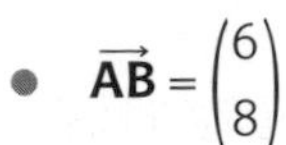

- $\overrightarrow{\mathbf{AB}} = \begin{pmatrix} 6 \\ 8 \end{pmatrix}$
- the length of the line segment AB can be found using Pythagoras's Theorem:

$$\text{AB}^2 = 6^2 + 8^2 = 100$$
$$\Rightarrow \quad \text{AB} = 10$$

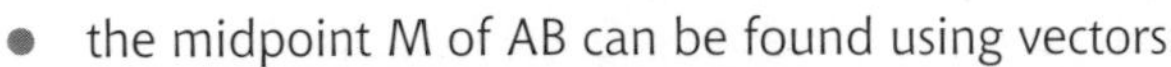

- the midpoint M of AB can be found using vectors:

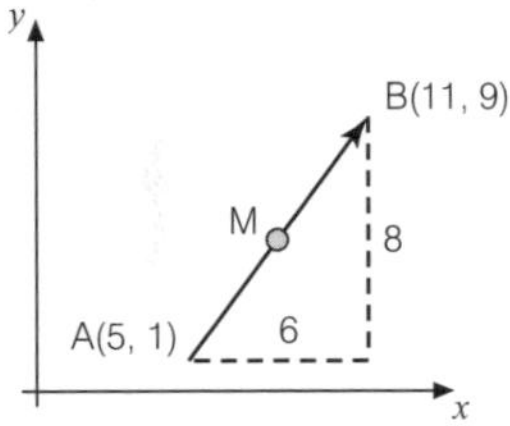

$$\overrightarrow{\mathbf{AM}} = \frac{1}{2}\overrightarrow{\mathbf{AB}} = \frac{1}{2}\begin{pmatrix} 6 \\ 8 \end{pmatrix} = \begin{pmatrix} 3 \\ 4 \end{pmatrix}$$

In moving from A(5, 1) to M the x co-ordinate changes by 3 so the x co-ordinate of M is 8.

In moving from A(5, 1) to M the y co-ordinate changes by 4 so the y co-ordinate of M is 5.

M is the point (8, 5).

These results can be generalised.

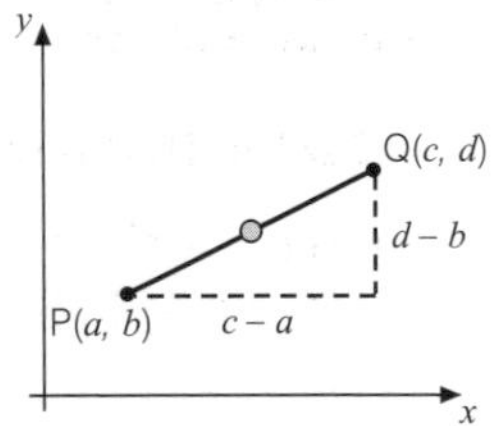

If P is the point (a, b) and Q is the point (c, d) then

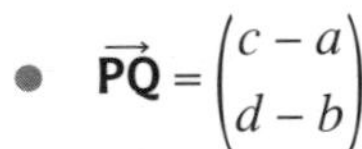

- $\overrightarrow{\mathbf{PQ}} = \begin{pmatrix} c-a \\ d-b \end{pmatrix}$
- the length of the line segment PQ can be found using Pythagoras's Theorem:

$$\text{PQ}^2 = (c-a)^2 + (d-b)^2$$
$$\text{so} \quad \text{PQ} = \sqrt{(c-a)^2 + (d-b)^2}$$

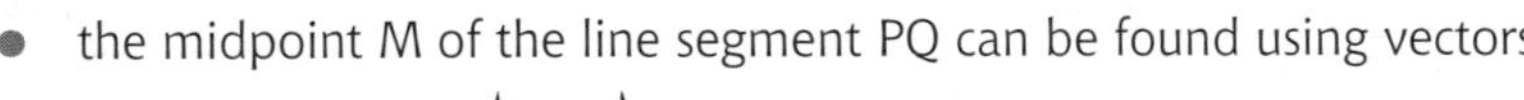

- the midpoint M of the line segment PQ can be found using vectors:

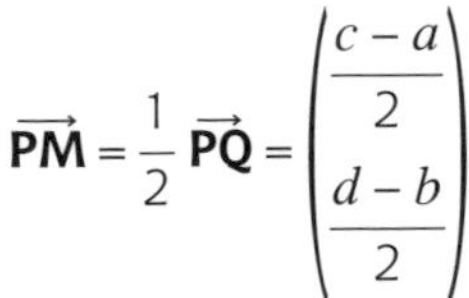

$$\overrightarrow{\mathbf{PM}} = \frac{1}{2}\overrightarrow{\mathbf{PQ}} = \begin{pmatrix} \dfrac{c-a}{2} \\ \dfrac{d-b}{2} \end{pmatrix}$$

In moving from P(a, b) to M the x co-ordinate changes by $\frac{c-a}{2}$ so

$$x \text{ co-ordinate of M} = a + \frac{c-a}{2} = \frac{2a+c-a}{2} = \frac{a+c}{2}.$$

Similarly, in moving from P(a, b) to M the y co-ordinate changes by $\frac{d-b}{2}$ so

$$y \text{ co-ordinate of M} = b + \frac{d-b}{2} = \frac{2b+d-b}{2} = \frac{b+d}{2}.$$

M is the point $\left(\frac{a+c}{2}, \frac{b+d}{2}\right)$.

To summarise:

If P is the point (a, b) and Q is the point (c, d) then

- distance $PQ = \sqrt{(c-a)^2 + (d-b)^2}$
- midpoint of the line segment joining P and Q is $M\left(\frac{a+c}{2}, \frac{b+d}{2}\right)$

EXAMPLE 1

P, Q and R are the points (−6, 2), (0, 10) and (2, −4), respectively.
Prove that the triangle PQR is isosceles.

SOLUTION

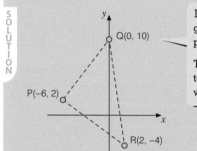

It is often worth drawing a sketch for co-ordinate geometry questions. Include the axes and make sure the points are placed approximately in the correct position.

These diagrams will often suggest possible approaches to the problems and will give a means of checking whether final answers are reasonable.

Using the formula for the distance between two points:

$$PQ = \sqrt{(0-(-6))^2 + (10-2)^2} = \sqrt{6^2 + 8^2} = \sqrt{100} = 10$$
$$QR = \sqrt{(2-0)^2 + ((-4)-10)^2} = \sqrt{2^2 + (-14)^2} = \sqrt{200}$$
$$PR = \sqrt{(2-(-6))^2 + ((-4)-2)^2} = \sqrt{8^2 + (-6)^2} = \sqrt{100} = 10$$

This is the **exact** length of QR.

You could write QR = 14.14 (2 d.p.) but this would only be an approximate answer.

PQ = PR so the triangle is isosceles.

If the points A and B lie on a horizontal or vertical line then the distance between the points or the midpoint can be written down without needing to use the distance formula.

For example, the distance between (1, 7) and (6, 7) is clearly 5 units and the midpoint of the line joining the two points is (3.5, 7).

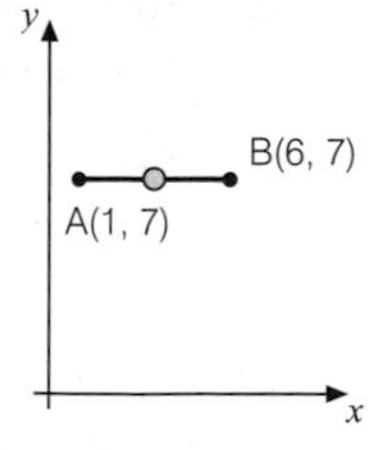

EXERCISE 1

1 **a)** If A is the point (3, 4) and $\overrightarrow{\mathbf{AB}} = \begin{pmatrix} 2 \\ 5 \end{pmatrix}$, write down the co-ordinates of B.

b) If C is the point (−1, 3) and $\overrightarrow{\mathbf{CD}} = \begin{pmatrix} -3 \\ 4 \end{pmatrix}$, write down the co-ordinates of D.

c) Find the vector $\overrightarrow{\mathbf{BD}}$.

2 **a)** If E is the point (−2, 4) and $\overrightarrow{\mathbf{EF}} = \begin{pmatrix} 7 \\ -5 \end{pmatrix}$, write down the co-ordinates of F.

b) If $\overrightarrow{\mathbf{FG}} = \begin{pmatrix} 2 \\ 4 \end{pmatrix}$, write down the co-ordinates of G.

c) Find the vector $\overrightarrow{\mathbf{EG}}$.
How is this answer connected to the vectors $\overrightarrow{\mathbf{EF}}$ and $\overrightarrow{\mathbf{FG}}$?
Draw a diagram to illustrate the points E, F and G and the vectors $\overrightarrow{\mathbf{EF}}$, $\overrightarrow{\mathbf{FG}}$ and $\overrightarrow{\mathbf{EG}}$.

3 If C is the point (6, 8) and D is the point (12, −4)
a) write down the vector $\overrightarrow{\mathbf{CD}}$
b) find the length of the line segment CD
c) write down the co-ordinates of the point M which is the midpoint of the line segment CD.

4 If P is the point (−3, 8) and Q is the point (2, −4)
a) write down the vector $\overrightarrow{\mathbf{PQ}}$
b) find the length of the line segment PQ
c) write down the co-ordinates of the point M which is the midpoint of the line segment PQ.

5 If U is the point (7, 3) and V is the point (13, −12) and W is the point on UV such that $\frac{UW}{UV} = \frac{1}{3}$
a) write down the vectors $\overrightarrow{\mathbf{UV}}$ and $\overrightarrow{\mathbf{UW}}$
b) find the co-ordinates of W.

6 Find the lengths of the line segments joining
a) (5, 3) to (8, 11)
b) (3, −2) to (−2, 1)
c) (7, 5) to (7, 17)
d) $(a, 3a)$ to $(4a, -a)$

7 Find the midpoints of the line segments joining
a) (6, 2) to (10, 12)
b) (−4, 7) to (1, 10)
c) (−6, 9) to (−6, −5)
d) $(-3p, 7p)$ to $(-7p, p)$

8 The triangle PQR has vertices P(−2, −3), Q(6, −2) and R(5, 1).
Find the exact lengths of PQ, QR and PR.
What can be deduced about the triangle PQR?

9 The triangle ABC has vertices A(6, 5), B(12, −3) and C(4, −1).
a) Calculate the length of the side AB.

M is the midpoint of the side AC and N is the midpoint of the side BC.
b) Find the co-ordinates of M and N.
c) Find the length of MN.

10 The triangle ABC has vertices A(5, 2), B(7, 9) and C(12, −2).
K, L and M are the midpoints of BC, AC and AB, respectively.
a) Write down the co-ordinates of the points K, L and M.

G is the point on the line AK such that $\frac{AG}{AK} = \frac{2}{3}$.
b) Write down the vectors $\overrightarrow{\mathbf{AK}}$ and $\overrightarrow{\mathbf{AG}}$ and hence find the co-ordinates of G.
c) Prove that $\overrightarrow{\mathbf{BG}} = \frac{2}{3}\overrightarrow{\mathbf{BL}}$. What does this tell you about the point G?
d) Prove that $\overrightarrow{\mathbf{CG}} = \frac{2}{3}\overrightarrow{\mathbf{CM}}$. What does this tell you about the point G?

The lines AK, BL and CM are called the medians of the triangle ABC. In this example we have seen that the medians all pass through the point G.
Try to generalise the argument of question 10 to show that the medians of **any** triangle pass through a single point G (called the centroid of the triangle) and that G is two thirds of the way from the vertex to the midpoint of the opposite side.

The Gradient of a Line Joining Two Points

The gradient of a line measures the steepness of the line, just as the gradient of a hill measures the steepness of a hill.

Suppose A is the point (1, 2) and B is the point (4, 6), then $\overrightarrow{\mathbf{AB}} = \begin{pmatrix} 3 \\ 4 \end{pmatrix}$ and

$$\text{gradient of AB} = \frac{y \text{ step}}{x \text{ step}} = \frac{4}{3}.$$

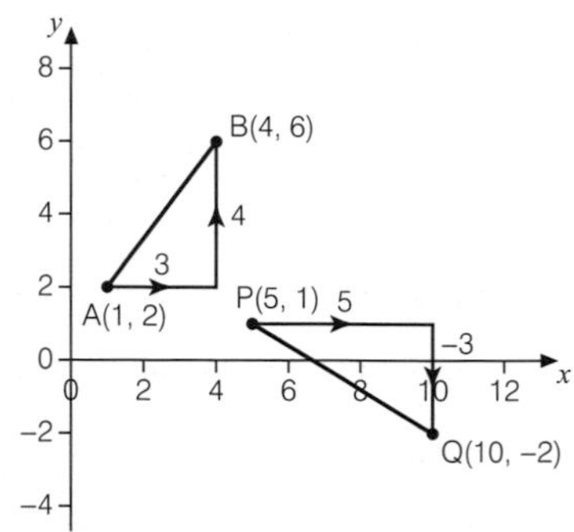

Similarly if P is the point (5, 1) and Q is the point (10, −2), then $\overrightarrow{\mathbf{PQ}} = \begin{pmatrix} 5 \\ -3 \end{pmatrix}$ and

$$\text{gradient of PQ} = \frac{y \text{ step}}{x \text{ step}} = \frac{-3}{5} = -\frac{3}{5}.$$

Notice that the gradient of the segment joining A to B is positive whilst the gradient of the segment joining P to Q is negative.

In general, as you move on the diagram from left to right along a line segment

- if you are moving up then the gradient is positive
- if you are moving down then the gradient is negative.

A formula for the gradient of the line joining two points can be derived:

If P is (a, b) and Q is (c, d) then

$$\overrightarrow{\mathbf{PQ}} = \begin{pmatrix} c-a \\ d-b \end{pmatrix}$$

y
Q(c, d)
P(a, b)
y step = d − b
x step = c − a
x

so

$$\text{gradient of PQ} = \frac{y \text{ step}}{x \text{ step}} = \frac{d-b}{c-a}.$$

The gradient of the line joining the points (a, b) and (c, d) is $\dfrac{d-b}{c-a}$

EXAMPLE 2

Find the gradient of the line segment joining the points (7, 5) and (11, −9)

SOL

$$\text{gradient} = = \frac{y \text{ step}}{x \text{ step}} = \frac{(-9)-5}{11-7} = \frac{-14}{4} = -3.5.$$

In the diagram below, the value, d, of the y co-ordinate of Q has been increased:

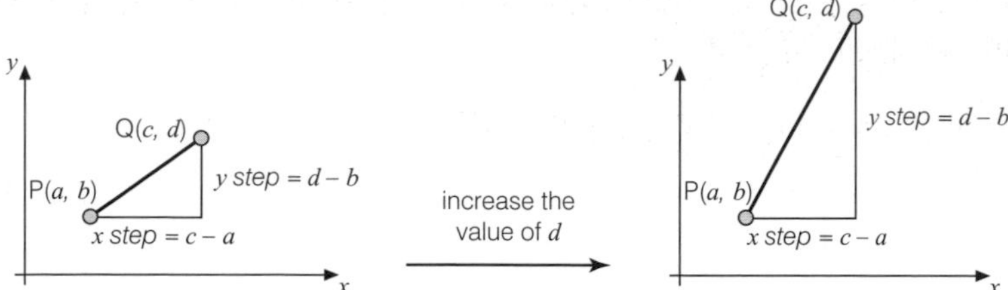

You can see that as d increases the line PQ becomes steeper. Moreover, the value of $d - b$ has increased but the value of $c - a$ has not changed, so the value of gradient PQ $= \dfrac{d-b}{c-a}$ has increased.

In the diagram below, the value, d, of the y co-ordinate of Q has been decreased:

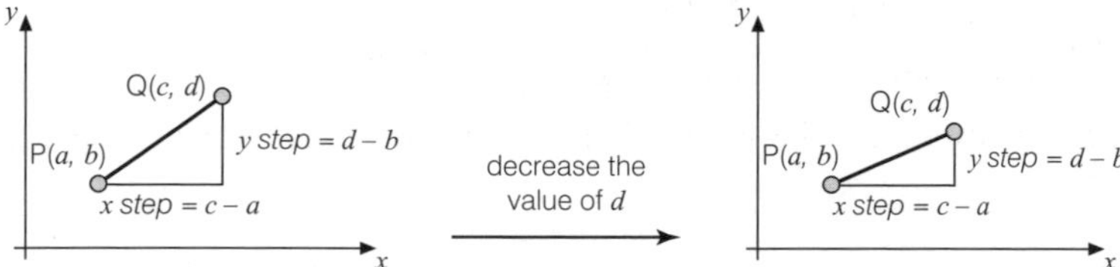

You can see that as d decreases the line PQ becomes less steep. Moreover, the value of $d - b$ has decreased but the value of $c - a$ has not changed, so the value of gradient PQ $= \dfrac{d-b}{c-a}$ has decreased.

These results confirm that:

- **the steeper a line is, the greater its gradient is.**

If the line segment joining two points is horizontal then the y step is zero and the gradient is therefore also zero:

- **a horizontal line segment has zero gradient.**

If the line segment joining two points is vertical then the x step is zero. Since division by 0 is not permitted, the gradient of a vertical line is not defined (we sometimes say that vertical lines have infinite gradient):

- **the gradient of a vertical line segment is not defined.**

The points A(–2, –3), B(0, 1), C(2, 5) and D(4, 9) all lie on a straight line.

Gradient AB $= \frac{4}{2} = 2$

Gradient AC $= \frac{6}{3} = 2$

Gradient AD $= \frac{12}{6} = 2$

Gradient BC $= \frac{2}{1} = 2$

Gradient BD $= \frac{8}{4} = 2$

Gradient CD $= \frac{6}{3} = 2$

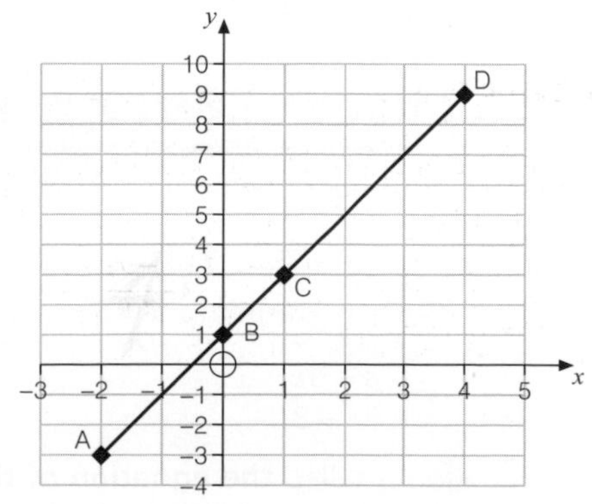

- **The gradients of all line segments on a straight line are the same.**

EXERCISE 2

1 **a)** If A is the point (4, 7) and B is the point (6, 16)
  i) write down the vector $\overrightarrow{\mathbf{AB}}$ **ii)** find the gradient of the line AB
b) If C is the point (8, −4) and D is the point (4, 4)
  i) write down the vector $\overrightarrow{\mathbf{CD}}$ **ii)** find the gradient of the line CD
c) If E is the point (−6, −3) and F is the point (4, −5)
  i) write down the vector $\overrightarrow{\mathbf{EF}}$ **ii)** find the gradient of the line EF

2 Calculate the gradient of the line segments joining the points
a) (2, 5) and (4, 13) **b)** (0, 0) and (5, 18) **c)** (5, 7) and (7, 6)
d) (−5, −7) and (7, 5) **e)** (−3, 8) and (5, 4) **f)** (4, −4) and (8, 12)
g) $(a, 3a)$ and $(4a, -6a)$ **h)** $(-2a, 3a)$ and $(4a, a)$ **i)** (a, a) and $(-2a, 4a)$.

3 Using scales of 1 cm : 1 unit for both axes, draw axes for $-8 \leqslant x \leqslant 8$ and $-8 \leqslant y \leqslant 8$.
a) If A, B, C and D are the points (3, 5), (4, 7), (1, −3) and (3, 1), respectively,
  i) draw the lines AB and CD
  ii) what can you say about these two lines?
  iii) find the gradient of AB and the gradient of CD
b) If E, F, G and H are the points (6, 4), (0, 6), (8, 3) and (−1, 6), respectively,
  i) draw the lines EF and GH
  ii) what can you say about these two lines?
  iii) find the gradient of EF and the gradient of GH

Write a sentence summarising what you have discovered in this question.

4 Using scales of 1 cm : 1 unit for both axes, draw axes for $-8 \leqslant x \leqslant 8$ and $-8 \leqslant y \leqslant 8$.
a) If A, B, C and D are the points (2, 1), (5, 7), (5, 1) and (1, 3), respectively,
  i) draw the lines AB and CD
  ii) what can you say about these two lines?
  iii) find the gradient of AB and the gradient of CD
b) If E, F, G and H are the points (−4, 8), (0, −4), (−2, 1) and (4, 3), respectively,
  i) draw the lines EF and GH
  ii) what can you say about these two lines?
  iii) find the gradient of EF and the gradient of GH
c) If J, K, L and M are the points (2, 6), (8, 0), (0, 1) and (6, 7), respectively,
  i) draw the lines JK and LM
  ii) what can you say about these two lines?
  iii) find the gradient of JK and LM. What do you notice?

Write a sentence summarising what you have discovered in this question.

The Equations of Straight Lines

A line passes through the points A(0, 3) and P(x, y) and has gradient 5.

$$\frac{y-3}{x} = \text{gradient AP} = 5$$

$$\Rightarrow \quad y - 3 = 5x$$

$$\Rightarrow \quad y = 5x + 3.$$

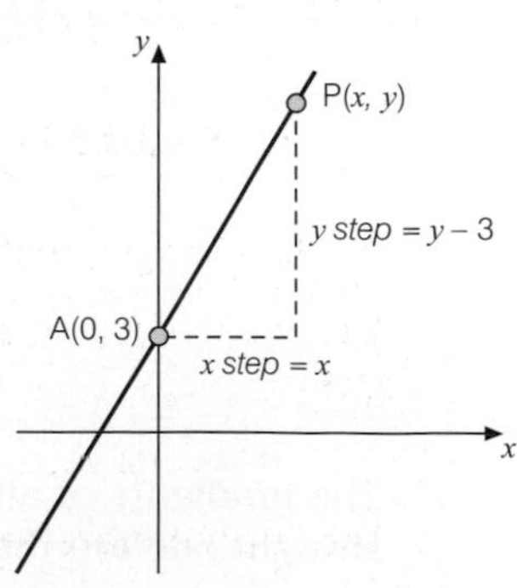

This equation is called **the equation of the straight line**.

Similarly, if a line passes through the points A(0, c) and P(x, y) and has gradient m, then

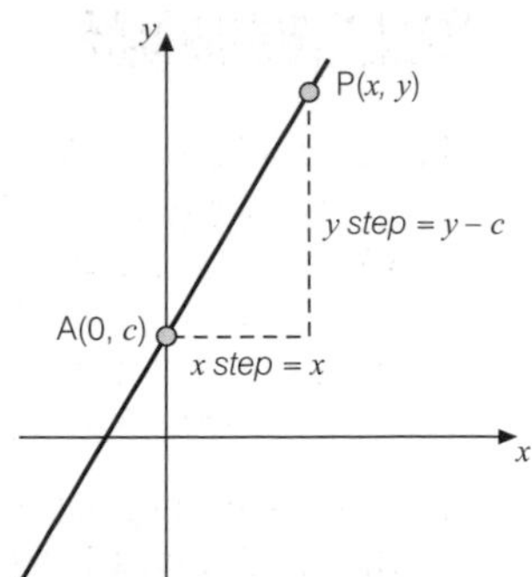

$$\frac{y-c}{x} = \text{gradient AP} = m$$

$$\Rightarrow \quad y - c = mx$$

$$\Rightarrow \quad y = mx + c.$$

This proves the important result:

> The straight line whose gradient is m and which crosses the y axis at the point (0, c) has equation
>
> $y = mx + c$

Note that the y co-ordinate of the point where the line crosses the y axis is sometimes called the y **intercept** of the line. It can be said that the equation $y = mx + c$ represents a line with gradient m and y intercept c.

You now know that

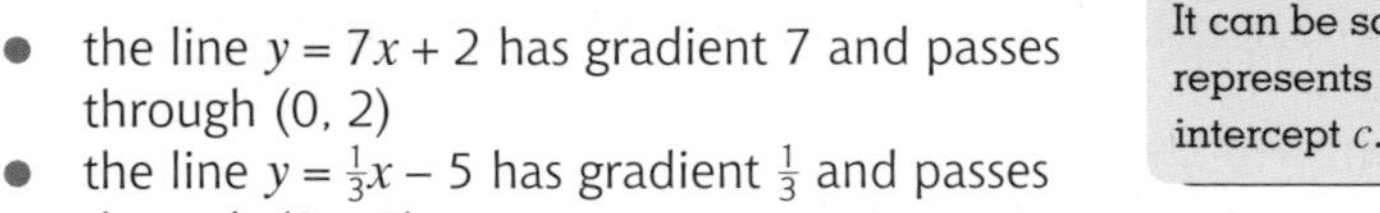

- the line $y = 7x + 2$ has gradient 7 and passes through (0, 2)
- the line $y = \frac{1}{3}x - 5$ has gradient $\frac{1}{3}$ and passes through (0, −5)
- the line $y = -3x + 2$ has gradient −3 and passes through (0, 2).

You also know that

- if a line has gradient 4 and passes through the point (0,6) then its equation must be $y = 4x + 6$
- if a line has gradient $\frac{2}{5}$ and passes through the point (0, −2) then its equation must be $y = \frac{2}{5}x - 2$.

EXAMPLE 3

Find the equation of the line through the point (5, 2) with gradient 3.

SOLUTION

The line has gradient 3 so its equation must be of the form $y = 3x + c$.

The line passes through the point (5, 2) so substituting $x = 5$, $y = 2$ into $y = 3x + c$ gives

$$2 = 15 + c \quad \Rightarrow \quad c = -13.$$

The line therefore has equation $y = 3x - 13$.

EXAMPLE 4

Find the equation of the line passing through the points A(4, 10) and B(8, −2).

SOLUTION

$$\text{Gradient AB} = \frac{y \text{ step}}{x \text{ step}} = \frac{-12}{4} = -3.$$

The line has gradient −3 so its equation must be of the form $y = -3x + c$.

The line passes through the point (4, 10) so substituting $x = 4$, $y = 10$ into $y = -3x + c$ gives

$$10 = -12 + c \quad \Rightarrow \quad c = 22.$$

The line therefore has equation $y = -3x + 22$.

EXERCISE 3

1 For each of the following lines write down the gradient and the co-ordinates of the point where the line crosses the y axis:

a) $y = 5x + 9$ **b)** $y = -4x + 11$ **c)** $y = -\frac{1}{2}x + 7$
d) $y = 2x - 7$ **e)** $y = -3x - 9$

2 Write down the equation of the line that

a) has gradient 6 and passes through (0, 14)
b) has gradient 2 and passes through (0, −5)
c) has gradient $-\frac{1}{4}$ and passes through (0, 1)
d) has gradient 12 and passes through (0, −11)

3 Find the equations of the line

a) through (3, 7) with gradient 4 **b)** through (−2, 5) with gradient 2
c) through (1, −2) with gradient −3 **d)** through (−2, −5) with gradient −2

4 If A is the point (5, 2) and B is the point (7, 8)

a) find the gradient of AB **b)** find the equation of the line AB

5 Find the equations of the line

a) through (1, 2) and (3, 8) **b)** through (−2, 8) and (2, 6)

Alternative Forms for the Equation of a Straight Line

The equation $y = mx + c$ is easy to use when the gradient and the y intercept are known. You have also seen that this equation can be adapted to deal with situations when the gradient and a point other than the y intercept are known. This situation occurs so frequently that it is worth having a second equation which you can use directly for such problems.

Consider the line of gradient m passing through the point $A(x_1, y_1)$.
Let $P(x, y)$ be another point on the line.

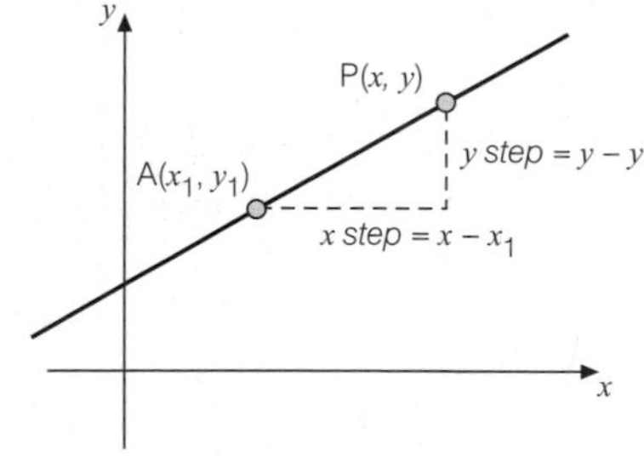

$$\text{Gradient AP} = \frac{y \text{ step}}{x \text{ step}} = \frac{y - y_1}{x - x_1} = m$$

$$\Rightarrow \quad y - y_1 = m(x - x_1).$$

So

> The equation
>
> $$y - y_1 = m(x - x_1)$$
>
> represents a line of gradient m through the point (x_1, y_1).

The equation of the line of gradient 7 through the point (5, −2) is therefore given by

$$y - (-2) = 7(x - 5)$$
$$\Rightarrow \quad y + 2 = 7x - 35$$
$$\Rightarrow \quad y = 7x - 37.$$

At times, the gradient and y intercept may be fractional values and, when this happens, the line equation in the $y = mx + c$ format can be somewhat unwieldy. In these circumstances, a modest amount of algebraic manipulation will allow you to rewrite the equation either as

$ax + by = c$ where a, b and c are integers

or as

$px + qy + r = 0$ where p, q and r are integers.

EXAMPLE 5

Find the equation of the line passing through the points A(3, 7) and B(9, 3), giving your answer in the form $ax + by = c$ where a, b and c are integers.

SOLUTION

Start by finding the gradient of AB:

$$\text{gradient AB} = \frac{y \text{ step}}{x \text{ step}} = \frac{-4}{6} = -\frac{2}{3}.$$

The line has gradient $-\frac{2}{3}$ and passes through (3, 7) so has equation

$$y - 7 = -\tfrac{2}{3}(x - 3)$$
$$\Rightarrow \quad y - 7 = -\tfrac{2}{3}x + 2$$
$$\Rightarrow \quad y = -\tfrac{2}{3}x + 9.$$

This is the "$y = mx + c$" answer.

Multiplying through by 3 gives

$$3y = -2x + 27$$
$$\Rightarrow \quad 3y + 2x = 27.$$

When the equation of a line is given in either $ax + by = c$ or $px + qy + r = 0$ format, the gradient and y intercept **cannot** be read immediately from the equation but can be deduced if some algebraic manipulation is used to convert the given format into the $y = mx + c$ format.

EXAMPLE 6

Show that the equation $5y - 2x + 7 = 0$ represents a straight line and find its gradient and y intercept.

SOLUTION

$$5y - 2x + 7 = 0$$
$$[-7] \quad \Rightarrow \quad 5y - 2x = -7$$
$$[+2x] \quad \Rightarrow \quad 5y = 2x - 7$$
$$[\div 5] \quad \Rightarrow \quad y = \tfrac{2}{5}x - \tfrac{7}{5}.$$

The equation $5y - 2x + 7 = 0$ is equivalent to the equation $y = \frac{2}{5}x - \frac{7}{5}$, so the equation represents a line of gradient $\frac{2}{5}$ and y intercept $-\frac{7}{5}$.

Observe that the gradient of $5y - 2x + 7 = 0$ is **not** -2 and the y intercept is **not** 7.

The gradient and y intercept should always be found by rearranging the equation into $y = mx + c$ format.

You have seen that one line can have many different, but equivalent, equations.

For example, the line $y = -\frac{1}{3}x + 6$ could also be written as $3y = -x + 18$ or $3y + x = 18$ or $3y + x - 18 = 0$.

Sometimes, a question will ask for the answer in a particular format and, when this is the case, it is important that your equation is in the correct format.

Sketching Straight Lines from their Equations

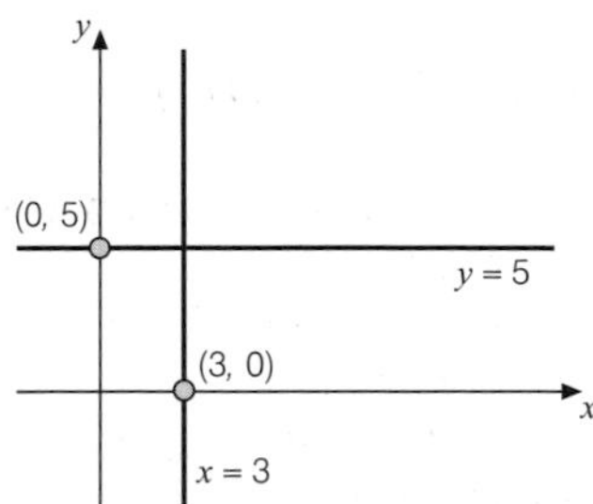

It is useful to be able to sketch straight lines from their equations.

The diagram shows sketches of the lines $x = 3$ and $y = 5$.

Remember that the line $x = a$ is a vertical line passing through $(a, 0)$. and that the line $y = b$ is a horizontal line passing through $(0, b)$.

The lines $y = x$ and $y = -x$ both pass through the origin and, if drawn accurately with the same scale on each axis, make an angle of 45° with the axes.

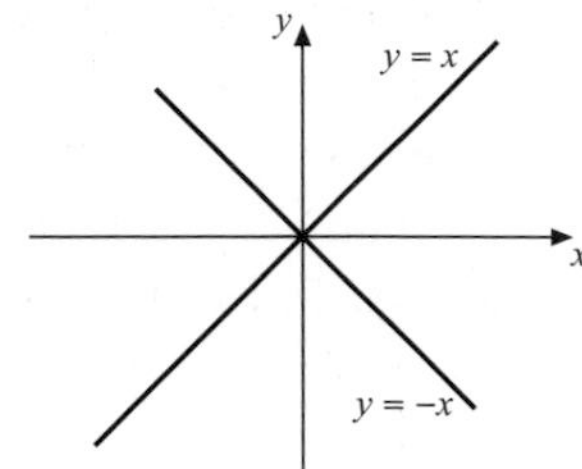

The line whose equation is $y = 2x + 3$ is a line with positive gradient passing through (0, 3).
The line whose equation is $y = 3.5x + 3$ is a steeper line also passing through (0, 3)
The line $y = -2x + 1$ is a line of negative gradient passing through (0, 1).

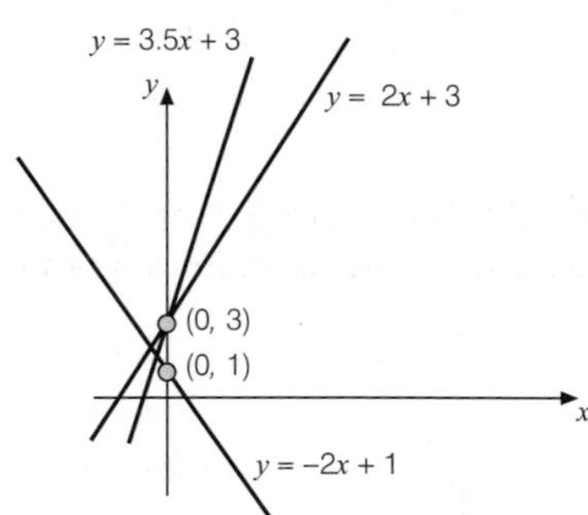

When the equation of a line is given in either $ax + by = c$ or $px + qy + r = 0$ format, the easiest way to sketch the graph is to find the two intercepts of the graph.

For example, to sketch $3x - 5y - 30 = 0$:
at the point where the line crosses the y axis, $x = 0$ so

$$3x - 5y - 30 = 0 \quad \Rightarrow \quad 0 - 5y - 30 = 0$$
$$\Rightarrow \quad 5y = -30$$
$$\Rightarrow \quad y = -6;$$

at the point where the line crosses the x axis, $y = 0$ so

$$3x - 5y - 30 = 0 \quad \Rightarrow 3x - 0 - 30 = 0$$
$$\Rightarrow \quad 3x = 30$$
$$\Rightarrow \quad x = 10.$$

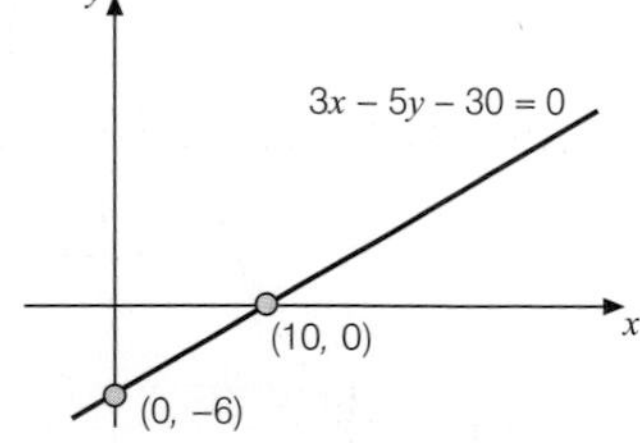

The line passes through (0, −6) and (10, 0). A sketch of the line is shown in the diagram.

In general, for any form of the straight line equation

- to find the y intercept, put $x = 0$ and solve for y
- to find the x intercept, put $y = 0$ and solve for x.

EXERCISE 4

1 **a)** Find the equation of the line with gradient 3 passing through (2, 11), giving your answer in $y = mx + c$ format.

b) Find the equation of the line passing through the points (5, 2) and (8, 6) giving your answer in the form $ax + by + c = 0$, where a, b and c are integers.

c) Find the equation of the line of gradient $-\frac{3}{4}$ passing through the point (−2, 7), giving your answer in the form $ax + by + c = 0$, where a, b and c are integers.

d) Find the equation of the line passing through the points (−2, 10) and (5, 6) giving your answer in the form $ax + by = c$, where a, b and c are integers.

2 Find the gradient and y intercept of the following lines:

a) $3x + 4y = 12$ **b)** $2x + y - 6 = 0$ **c)** $y = 2x + 5$

d) $2y - 5x - 8 = 0$ **e)** $7y + 2x - 14 = 0$ **f)** $6 + 2y = 5x$

g) $3x = 2y - 6$ **h)** $y = -4x + 7$ **i)** $4y = 3x + 6$

3 Sketch, on separate diagrams, the graphs of

a) $y = 2$ **b)** $x = -6$ **c)** $y = 3x$

d) $y = -\frac{1}{2}x$ **e)** $y = 2x - 3$ **f)** $y = 5x - 3$

g) $y = -2x - 3$ **h)** $2x + 7y - 14 = 0$ **i)** $5x - 4y = 40$

j) $2x - 5y - 20 = 0$ **k)** $7x - 2y = 0$ **l)** $3x - 4y = 10$

Parallel and Perpendicular Lines

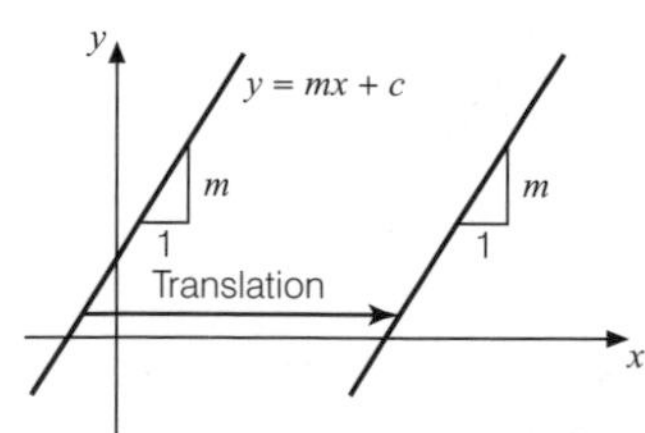

If the line $y = mx + c$ is translated then the image is a parallel line and the gradient of the new line is the same as the gradient of the original line.

Parallel lines have equal gradients.

Lines that meet at right angles are called **perpendicular** lines.

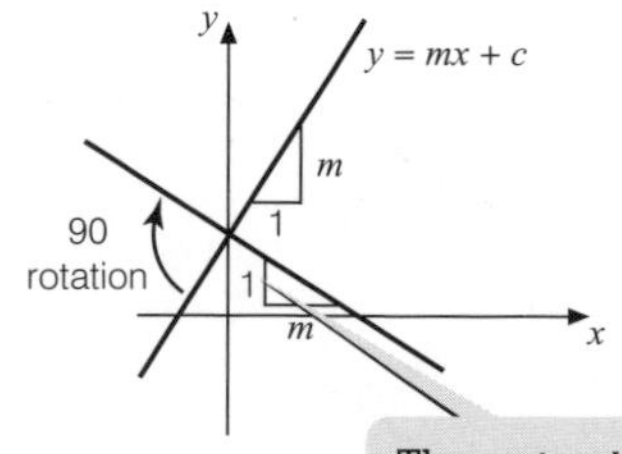

If the line $y = mx + c$ is rotated by 90° about a point on the line then a perpendicular line is obtained.

$$\text{Gradient of new line} = \frac{y\text{ step}}{x\text{ step}} = \frac{-1}{m} = -\frac{1}{m}.$$

The y-step is −1 here because we have to move 1 unit down.

Thus

$$\text{gradient of original line} \times \text{gradient of perpendicular line} = m \times \left(-\frac{1}{m}\right) = -1.$$

The product of the gradients of perpendicular lines is −1.

This important result is summarised by saying:

If lines with gradients m_1 and m_2 are perpendicular then $m_1 m_2 = -1$.

EXAMPLE 7

Find the equation of the line

a) which passes through (2, −5) and is parallel to the line $y = 4x - 3$
b) which passes through (2, −5) and is perpendicular to the line $y = 4x - 3$.

SOLUTION

a) The line $4x - 3$ has gradient 4 so the parallel line must also have gradient 4. The parallel line therefore has equation

$$y = 4x + c.$$

The parallel line passes through (2, −5) so

$$-5 = 4 \times 2 + c$$
$$\Rightarrow \quad c = -13.$$

The parallel line has equation

$$y = 4x - 13.$$

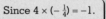

Since $4 \times (-\frac{1}{4}) = -1$.

b) The line $y = 4x - 3$ has gradient 4 so the perpendicular line must have gradient $-\frac{1}{4}$. The perpendicular line therefore has equation

$$y = -\tfrac{1}{4}x + c.$$

The perpendicular line passes through (2, −5) so

$$-5 = -\tfrac{1}{4} \times 2 + c$$
$$\Rightarrow \quad c = -\tfrac{9}{2}.$$

The perpendicular line has equation

$$y = -\tfrac{1}{4}x - \tfrac{9}{2}$$

or, equivalently

$$4y + x = -18.$$

EXAMPLE 8

Find the equation of the perpendicular bisector of the line joining the points C(5, 4) and D(9, −2) giving your answer in the form $ax + by + c = 0$ where a, b and c are integers.

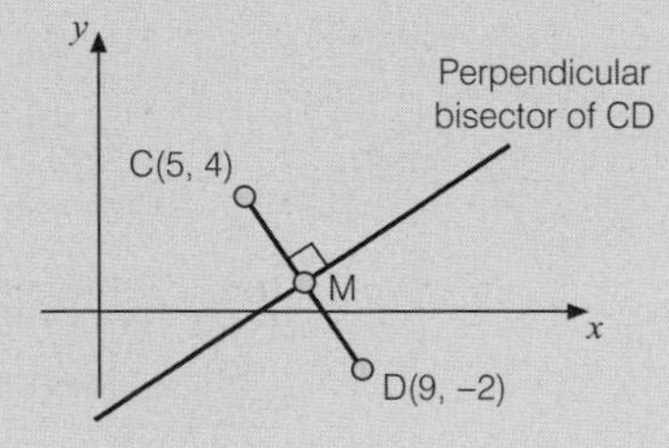

SOL

The perpendicular bisector of CD passes through the midpoint M of the line joining C to D and is at right angles to the line CD.

EXAMPLE 8 (continued)

Since C is (5, 4) and D is (9, −2), M must be $\left(\frac{5+9}{2}, \frac{4+(-2)}{2}\right) \Rightarrow M = (7, 1)$.

Gradient of CD $= \frac{-6}{4} = -\frac{3}{2}$. The perpendicular line therefore has gradient $\frac{2}{3}$.

Since $(-\frac{3}{2}) \times (\frac{2}{3}) = -1$.

The perpendicular bisector has gradient $\frac{2}{3}$ and passes through M(7, 1) so its equation is

$$y - 1 = \tfrac{2}{3}(x - 7)$$
$$\Rightarrow \quad 3y - 3 = 2(x - 7)$$
$$\Rightarrow \quad 3y - 3 = 2x - 14$$
$$\Rightarrow \quad 3y - 2x + 11 = 0.$$

You want the final answer in $ax + by + c = 0$ format so it is worth multiplying throughout by 3 to remove the fractions before expanding.

EXERCISE 5

1 If A, B, P and Q are the points (2, 5), (−2, 9), (5, 0) and (0, 5), respectively,
a) find the gradient of AB and PQ
b) what can you deduce about these two lines?

2 If A, B, P and Q are the points (1, 3), (3, 2), (−1, 5) and (1, 9), respectively,
a) find the gradient of AB and PQ
b) what can you deduce about these two lines?

3 If A, B, P and Q are the points $(p, 3p)$, $(2p, 6p)$, $(3p, -p)$ and $(6p, -2p)$, respectively,
a) find the gradient of AB and PQ
b) what can you deduce about these two lines?

4 If A, B and C are the points (5, 0), (13, 6) and (8, −4), respectively, prove that AB and AC are perpendicular. Find the lengths of AB and AC and hence find the area of the triangle ABC.

5 The equations of six lines are given below. Determine which lines are parallel to each other and which lines are perpendicular to each other.

Line 1	$y = 5x - 4$	Line 2	$y = -5x + 3$
Line 3	$y = \frac{1}{5}x + 2$	Line 4	$y - 5x = 10$
Line 5	$x - 5y = 8$	Line 6	$y = \frac{-1}{5}x + 3$

6 Find the equation of the line
a) through (3, −2) and parallel to $y = 2x + 8$, giving your answer in the form $y = mx + c$
b) through (4, −3) and perpendicular to $y = 4x - 2$, giving your answer in the form $ax + by + c = 0$ where a, b and c are integers
c) through (3, 2) and parallel to $y = -4x$, giving your answer in the form $y = mx + c$
d) through (8, 2) and perpendicular to $y + 4x = 8$, giving your answer in the form $ax + by = c$ where a, b and c are integers
e) through (1, 5) and perpendicular to $2x + y - 12 = 0$, giving your answer in the form $y = mx + c$.

7 Find the equation of the perpendicular bisector of the line segment joining the points (−2, 6) and (8, 4) giving your answer in the form $ax + by = c$ where a, b and c are integers.

8 The image of the point (7, 11) after a reflection is the point (13, 3). Find the equation of the mirror line.

9 The diagram shows three lines, L_1, L_2 and L_3. L_1 has equation $y = 2x - 6$.

a) Find the co-ordinates of the points A and B.

The line L_2 is parallel to L_1 and passes through the point (8, 0).

b) Find the equation of L_2 and the co-ordinates of the point where L_2 meets the y-axis.

The line L_3 is perpendicular to L_1 and passes through (0, 6).

c) Find the co-ordinates of the point where L_3 meets the x-axis.

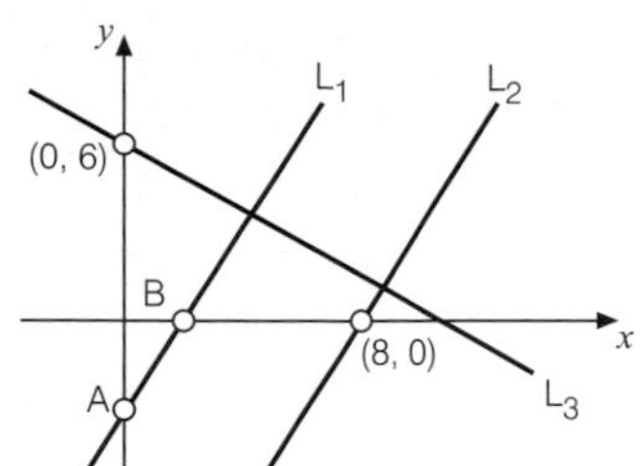

Using Simultaneous Equations to Find Points of Intersection

If two lines are not parallel then they will meet in a unique point called the point of intersection of the two lines. The point (x, y), where the lines cross, lies on both of the lines so its co-ordinates satisfy both of the straight line equations. This means that finding the point of intersection involves solving a pair of simultaneous equations.

EXAMPLE 9

Find the point of intersection of the lines $3x + 2y = 11$ and $2x - 3y = 16$.

SOLUTION

You must solve the simultaneous equations:

$$\left.\begin{aligned} 3x + 2y &= 11 \\ 2x - 3y &= 16 \end{aligned}\right\}$$

$$3x + 2y = 11 \xrightarrow{\times 2} 6x + 4y = 22$$

$$2x - 3y = 16 \xrightarrow{\times 3} 6x - 9y = 48$$

$$[-] \quad 13y = -26$$

$$\Rightarrow \quad y = -2.$$

Substitute this in the first equation:

$$3x - 4 = 11$$

$$\Rightarrow \quad 3x = 15$$

$$\Rightarrow \quad x = 5.$$

The lines intersect at the point (5, −2).

This is probably the format for simultaneous equations that you are already familiar with. The equations can generally be solved by doing the following.

- Multiply up the equations by suitable numbers to balance the coefficients of one of the two variables.
- Eliminate the variable with the balanced coefficients by adding or subtracting as appropriate.
- Solve the resulting equation to find the value of one of the variables.
- Substitute this value back into one of the original equations and solve the resulting equation to find the value of the second variable.

EXAMPLE 10

Find the point of intersection of the lines $3x + 2y = 6$ and $y = 2x - 11$.

SOLUTION

You must solve the simultaneous equations:

$$\left.\begin{aligned} 3x + 2y &= 6 \\ y &= 2x - 11 \end{aligned}\right\}$$

$$\left.\begin{aligned} 3x + 2y &= 6 \\ y &= 2x - 11 \end{aligned}\right\} \Rightarrow 3x + 2(2x - 11) = 6$$

$$\Rightarrow 7x - 22 = 6$$
$$\Rightarrow 7x = 28$$
$$\Rightarrow x = 4$$
$$\Rightarrow y = 2 \times 4 - 11 = -3.$$

The lines intersect at the point $(4, -3)$.

> These equations could be solved by rewriting the second equation as
>
> $-2x + y = -11$
>
> and then proceeding in a similar way to example 9.
>
> However it is quicker in this case to use the second equation to rewrite the first equation with no mention of y.

EXAMPLE 11

Find the point of intersection of the lines $y = 10 - x$ and $y = 2x - 11$.

SOLUTION

You must solve the simultaneous equations.

$$\left.\begin{aligned} y &= 10 - x \\ y &= 2x - 11 \end{aligned}\right\}$$

$$\left.\begin{aligned} y &= 10 - x \\ y &= 2x - 11 \end{aligned}\right\} \Rightarrow 10 - x = 2x - 11$$

$$\Rightarrow 21 = 3x$$
$$\Rightarrow x = 7$$
$$\Rightarrow y = 10 - 7 = 3.$$

The lines intersect at the point $(7, 3)$.

> Again, the equations could be solved by converting them to the form of example 9.
>
> However it is much quicker to simply equate the two different expressions for y.

EXERCISE 6

> Your graphical calculator may well solve simultaneous equations for you. If this is the case, make sure you use your calculator to check your solutions.
> In future modules you will be able to use your calculator to solve the equations **but in this module you must be able to solve simultaneous equations without using a calculator.**

Find the points of intersection of the following pairs of lines:

1. $2x + 5y = 3$
   $3x - 2y = 14$
2. $y = 2x - 5$
   $y = x - 1$
3. $x + 2y = 1$
   $y = 2x - 12$
4. $y = 3x - 5$
   $3x + 2y = 8$
5. $4x - 3y = 11$
   $3x - 5y = 11$
6. $y = 7 - x$
   $y = 2x + 1$
7. $x + y = 9$
   $3x - 5y = 11$
8. $x = 2y - 9$
   $x = y - 4$
9. $x = 2y + 1$
   $3x - 4y = 7$
10. $x = 2y - 7$
    $y = 2x - 1$

11 The points A, B, C and D are (2, −2), (4, −3), (6, −4) and (7, −2), respectively.
- **a)** Prove that AB is perpendicular to CD and that the line segments AB and CD have the same length.
- **b)** Prove that the perpendicular bisector of the line segment joining the points A and C has equation $y = 2x - 11$.
- **c)** Obtain the equation of the perpendicular bisector of the line segment joining the points B and D.
- **d)** Find the point of intersection of these two perpendicular bisectors.
- **e)** Hence describe fully the transformation that maps A to C and B to D.

12 The triangle ABC has vertices A(0, 4), B(6, 7) and C(4, 1).
- **a)** Prove that the line L_1 that is perpendicular to the line segment AB and passes through C has equation $2x + y = 9$.
- **b)** Find the equation of the line L_2 which is perpendicular to the line segment AC and passes through B. Give your answer in the form $ax + by = c$, where a, b and c are integers.
- **c)** The point P is the point of intersection of L_1 and L_2. Obtain the co-ordinates of P.
- **d)** Prove that the line segments AP and BC are perpendicular.

Areas of Shapes

Areas of complicated shapes may be calculated by a simple boxing off procedure.

EXAMPLE 12

Find the area of the quadrilateral with vertices A(3, 1), B(5, 2), C(1, 3) and D(−2, 2).

SOLUTION

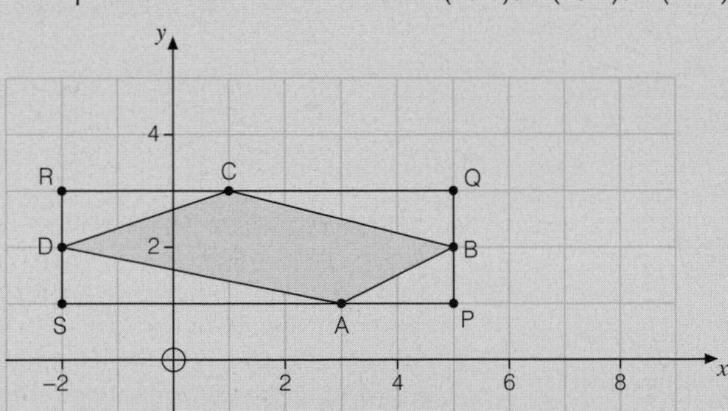

The quadrilateral has been enclosed within a rectangle whose sides are parallel to the axes.

$$\begin{aligned}\text{Area of quadrilateral} &= \text{area of PQRS} - \\ &\quad (\text{area } \Delta\text{APB} + \text{area } \Delta\text{BQC} + \text{area } \Delta\text{CRD} + \text{area } \Delta\text{DSA}) \\ &= 7 \times 2 - (\tfrac{1}{2} \times 2 \times 1 + \tfrac{1}{2} \times 1 \times 4 + \tfrac{1}{2} \times 3 \times 1 + \tfrac{1}{2} \times 1 \times 5) \\ &= 14 - (1 + 2 + \tfrac{3}{2} + \tfrac{5}{2}) \\ &= 14 - 7 \\ &= 7.\end{aligned}$$

EXAMPLE 13

The vertex A of a parallelogram ABCD has co-ordinates (3, 1). The equation of BC is $y = x + 4$, and the equation of the diagonal BD is $2y = 5x - 4$. Calculate the co-ordinates of the vertices B, C and D and find the area of the parallelogram.

SOLUTION

A good sketch diagram to show the information given is often an appropriate start to complicated questions like this. Try to ensure that the lines and points are roughly in the right place.

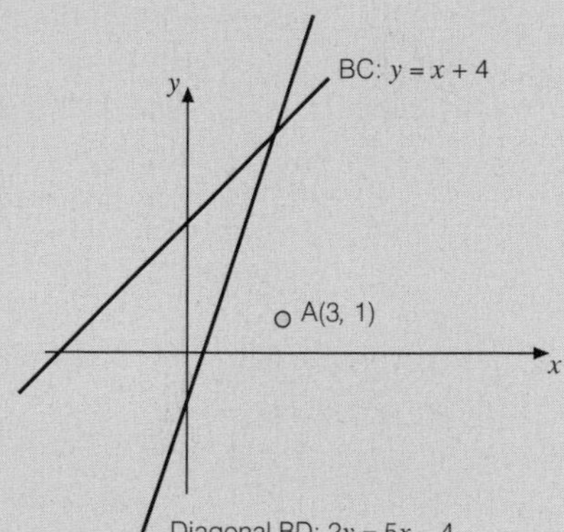

This first diagram shows the information given in the question.

You can improve the information in the diagram by

- including the position of B on the diagram since it is the point of intersection of the lines BC and BD which are already shown
- then drawing the side AB of the parallelogram
- using the fact that the side AD of the parallelogram passes through A and must be parallel to the opposite side BC to add the side AD
- marking the point D as the point of intersection of the side AD and the diagonal BD
- placing C using the fact that the movement from D to C must be the same as the movement from A to B.

This gives

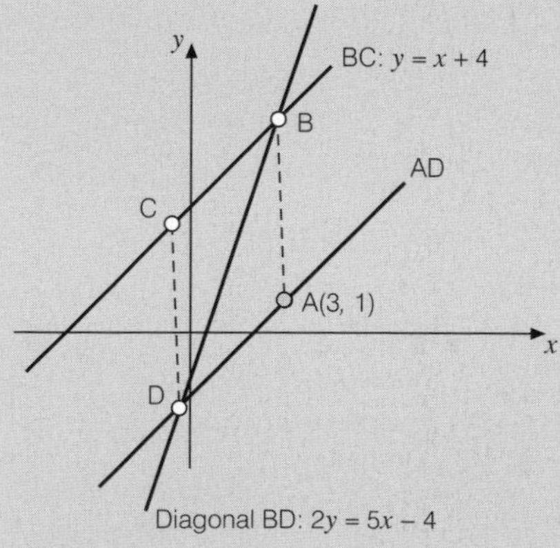

This diagram is certainly not accurate but it gives you an idea of what the parallelogram looks like.
Moreover, the steps of drawing the diagram give a strategy for answering the question.

B is the intersection of $y = x + 4$ and $2y = 5x - 4$:

$$\left.\begin{array}{r} y = x + 4 \\ 2y = 5x - 4 \end{array}\right\} \quad \Rightarrow \quad 2(x + 4) = 5x - 4$$

$$\Rightarrow \quad 2x + 8 = 5x - 4$$

$$\Rightarrow \quad 12 = 3x$$

$$\Rightarrow \quad x = 4 \qquad \text{and} \qquad y = x + 4 \quad \Rightarrow \quad y = 8.$$

So B is (4, 8).

EXAMPLE 13 (continued)

The line BC has gradient 1; the line AD is parallel to BC so must have the same gradient.
The line AD has gradient 1 and passes through (3, 1) so its equation is

$$y - 1 = 1(x - 3)$$
$$\Rightarrow \quad y - 1 = x - 3$$
$$\Rightarrow \quad y = x - 2.$$

The point D is the point of intersection of $y = x - 2$ and $2y = 5x - 4$:

$$\left.\begin{array}{r} y = x - 2 \\ 2y = 5x - 4 \end{array}\right\} \Rightarrow \quad 2(x-2) = 5x - 4$$
$$\Rightarrow \quad 2x - 4 = 5x - 4$$
$$\Rightarrow \quad 0 = 3x$$
$$\Rightarrow \quad x = 0 \quad \text{and} \quad y = x - 2 \quad \Rightarrow \quad y = -2.$$

So D is (0, −2).

The point C can easily be found using a vector argument: $\overrightarrow{\mathbf{BC}} = \overrightarrow{\mathbf{AD}} = \begin{pmatrix} -3 \\ -3 \end{pmatrix}$ and you know that B is the point (4, 8) so C must be (1, 5).

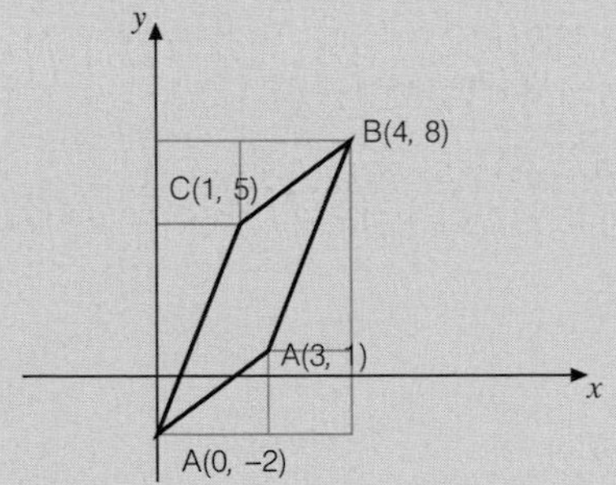

The area of the parallelogram can now be found by drawing a rather more accurate sketch and then boxing off:

$$\text{Area} = 4 \times 10 - [\tfrac{1}{2} \times 3 \times 3 + 3 \times 1 + \tfrac{1}{2} \times 1 \times 7] \times 2 = 18 \text{ units}^2.$$

EXERCISE 7

1 Find the area of

a) the triangle with vertices (4, 2), (7, 3) and (−1, 7)

b) the quadrilateral with vertices (3, 1), (8, 4), (4, 7) and (2, 4)

c) the pentagon with vertices (−1, −3), (4, −1), (5, 2), (1, 5) and (−2, 1)

2 The line $3x + 4y = 12$ crosses the x axis at P and the y axis at Q. The line $5x + 3y = 30$ crosses the y axis at R and the x axis at S. Calculate the area of the quadrilateral PQRS.

3 The points A(0, −2) and B(4, 4) are adjacent vertices of the rectangle ABCD and the point D lies on the x axis.

i) Determine the equation of the line AD and deduce the co-ordinates of D.

ii) Find the co-ordinates of C.

iii) Calculate the area of the rectangle.

4 The points A(–1, 2) and C(5, 1) are opposite vertices of a parallelogram ABCD. The vertex B lies on the line $2x + y = 5$. The side AB is parallel to the line $3x + 4y = 8$. Calculate

i) the equation of the side AB
ii) the co-ordinates of B;
iii) the equations of the sides AD and CD
iv) the co-ordinates of D
v) the area of the parallelogram

5 *(Recall that a rhombus is a parallelogram with all four sides equal in length and that the diagonals of a rhombus bisect each other at right angles.)*

A rhombus PQRS has P(9, 6) and the centre of the rhombus is W(11, 2).

a) Determine the co-ordinates of R.
b) Find the equation of QS.

The line PQ has equation $4x - 3y - 18 = 0$.

c) Determine the co-ordinates of Q and S.
d) Prove that the perimeter of the rhombus is 40 units long and determine its area.

6 The triangle ABC has vertex A on the x axis and vertex B on the y axis. C is the point (11, 5) and the foot, F, of the perpendicular from C to AB is the point (9, 2).

a) Find the equation of the line AB.
b) Write down the co-ordinates of the points A and B.
c) Calculate the area of the triangle ABC.

7 If A, B and C are the points (–2, 3), (3, 3) and (1, 7), respectively,

i) prove that AB and AC have the same length;
ii) find the equation of the line, L, which is the perpendicular bisector of BC;
iii) show that A lies on L.

What relationship does the line L have to angle BAC?

Having studied this chapter you should know that

- if P is the point (a, b) and Q is the point (c, d) then

  distance PQ $= \sqrt{(c-a)^2 + (d-b)^2}$

  gradient of PQ $= \dfrac{d-b}{c-a}$

  midpoint of PQ is $M\left(\dfrac{a+c}{2}, \dfrac{b+d}{2}\right)$
- the equation $y = mx + c$ represents a straight line of gradient m and y intercept c
- the equation $y - y_1 = m(x - x_1)$ represents a straight line of gradient m passing through the point (x_1, y_1)
- equations of the form $ax + by = c$ and $ax + by + c = 0$ also represent straight lines. Their gradients and y intercepts can be found by rearranging the equation into "$y = mx + c$" format
- parallel lines have equal gradients
- if $y = m_1x + c_1$ and $y = m_2x + c_2$ are perpendicular then $m_1m_2 = -1$
- the point of intersection of two lines may be found by solving their equations simultaneously
- areas of polygons can be found by a boxing off process

REVISION EXERCISE

1 If A, B and C are the points (5, 5), (−1, 3) and (3, −1), respectively, find the lengths AB and AC.
What can you say about the triangle ABC?

2 The co-ordinates of the points A and B are (2, 6) and (−3, 4), respectively.
i) Calculate the gradient of AB.
ii) Find the equation of the straight line which passes through A and is perpendicular to AB. Give your answer in the form $px + qy = r$.

(OCR Nov 1997 P1)

3 **a)** Write down the co-ordinates of the midpoints of the line segments joining
i) (2, −1) and (2, 5) **ii)** (−3, 0) and (−13, 8)

b) If A is the point (−4, 8) and B is the point (1, −2) find the co-ordinates of the point T on the line segment AB which satisfies $\dfrac{\text{AT}}{\text{TB}} = \dfrac{1}{4}$.

4 Find the gradient of the line $2y + 5x - 10 = 0$.
Find the equation of the line which passes through the point (3, 4) that is parallel to the line $2y + 5x - 10 = 0$. Give your answer in the form $ax + by + c = 0$, where a, b and c are integers. Find also the equation of the line which passes through the point (3, 4) that is perpendicular to the line $2y + 5x - 10 = 0$, giving your answer in the form $ax + by + c = 0$, where a, b and c are integers.

5 Find the equation of the line joining the points A(6, 3) and B(5, 8).
Find the equation of the line that is perpendicular to this line and passes through the point A. Verify that the point C(1, 2) lies on this line.
Find the area of the triangle ABC.

6 Find the gradient of the line $x + 4y - 1 = 0$.
Find the point of intersection of the lines $y = 2x$ and $x + y - 3 = 0$.
Find the equation of the line that is parallel to the line $x + 4y - 1 = 0$ and which passes through the point of intersection of the lines $y = 2x$ and $x + y - 3 = 0$. Give your answer in the form $ax + by + c = 0$ where a, b and c are integers.

7 Two sides, AB and AD, of the parallelogram ABCD have equations $7y - x + 13 = 0$ and $9x - y + 7 = 0$, respectively, and the centre of the parallelogram is the point P(3, 3). Calculate the co-ordinates of the four vertices A, B, C and D.

8 If P, Q and R are the points (4, 6), (6, 2) and (5, −1), respectively,

a) find the equation of the line L_1 which is the perpendicular bisector of PQ
b) find the equation of the line L_2 which is the perpendicular bisector of PR
c) prove that the point of intersection of L_1 and L_2 is the point C(1, 2)
d) calculate the distances CP, CQ and CR
e) what can you say about the circle whose centre is C and whose radius is 5?

9 a) The point A has co-ordinates (2, 3) and the line L_1 has equation $x + 4y = 31$. The line L_2 passes through A and is perpendicular to L_1. Find the equation of L_2 in the form $y = mx + c$.
b) The lines L_1 and L_2 intersect at the point M. Find the co-ordinates of M.
c) The point A is the vertex of the square ABCD. The diagonals of the square intersect at M. Find the co-ordinates of C.

(OCR Jan 2002 P1)

10 The point A has co-ordinates (1, 7) and the point B has co-ordinates (3, 1). The midpoint of AB is P. Find the equation of the line which passes through P and which is perpendicular to the line $5y + x = 7$. Give your answer in the form $y = mx + c$.

(OCR Jun 2001 P1)

11 The straight line p passes through the point (10, 1) and is perpendicular to the line r with equation $2x + y = 1$. Find the equation of p.
Find also the co-ordinates of the point of intersection of p and r and deduce the perpendicular distance from the point (10, 1) to the line r.

(OCR Jun 1995 P1)

12 The equation of the straight line L_1 is $x + 3y - 33 = 0$. The point P is (3, 0) and the point Q is (6, 9). The straight line L_2 is parallel to L_1 and passes through P.

i) Find the equation of L_2, giving your answer in the form $ax + by + c = 0$.
ii) Verify that Q lies on L_1.
iii) Show that the line joining P to Q is perpendicular to L_1.
iv) Find the perpendicular distance between L_1 and L_2.

(OCR May 1996 P1)

5 Surds

The purpose of this chapter is to enable you to

- use surds, where appropriate, to give exact answers
- use the properties of surds to simplify expressions involving surds

It was demonstrated in the previous chapter that the distance between the points P(a, b) and Q(c, d) can be calculated using the formula

$$PQ = \sqrt{(c-a)^2 + (d-b)^2}$$

and that, in many cases, the only way of writing down the exact distance between the two points is to leave a square root in the expression.
For example the distance between the points (0, 10) and (2, −4) is

$$\sqrt{(2-0)^2 + (-4-10)^2} = \sqrt{4+196} = \sqrt{200}.$$

The value $\sqrt{200}$ is an **exact** answer for the distance between the two points.
If a calculator is now used to obtain 14.14 (to two decimal places) then the answer is only an approximation to the distance between the two points.

The number $\sqrt{200}$ cannot be simplified to a rational number.

A rational number is any number that can be written as a fraction of two integers.

$\sqrt{200}$ is an example of a surd: any root that cannot be simplified to a rational number is a surd. When an "exact" answer is given in a form involving a root that cannot be simplified to a rational number, the answer is said to be given in **surd form**.

In this chapter the arithmetic of surds is developed to enable you to give exact answers in their simplest possible form.

Simplifying Surd Expressions

Recall from chapter 1 that

$$p^{\frac{1}{2}} = \sqrt{p}$$

and that

$$(pq)^n = p^n q^n$$

$$\left(\frac{p}{q}\right)^n = \frac{p^n}{q^n}.$$

Putting $n = \frac{1}{2}$ into these two results gives

$$(pq)^{\frac{1}{2}} = p^{\frac{1}{2}} q^{\frac{1}{2}} \quad \Rightarrow \quad \sqrt{pq} = \sqrt{p}\sqrt{q}$$

$$\left(\frac{p}{q}\right)^n = \frac{p^n}{q^n} \quad \Rightarrow \quad \sqrt{\frac{p}{q}} = \frac{\sqrt{p}}{\sqrt{q}}.$$

These two results provide the key to simplifying many expressions involving surds:

$$\sqrt{pq} = \sqrt{p}\sqrt{q} \qquad \sqrt{\frac{p}{q}} = \frac{\sqrt{p}}{\sqrt{q}}$$

Using these results you can write:

$$\sqrt{50} = \sqrt{25 \times 2} = \sqrt{25} \times \sqrt{2} = 5\sqrt{2}$$

$$\begin{aligned}\sqrt{27} + \sqrt{32} &= \sqrt{9 \times 3} + \sqrt{16 \times 2}\\ &= 3\sqrt{3} + 4\sqrt{2}\end{aligned}$$

$$\begin{aligned}\sqrt{48} - \sqrt{27} &= \sqrt{4 \times 12} - \sqrt{9 \times 3}\\ &= 2\sqrt{12} - 3\sqrt{3}\\ &= 2\sqrt{4 \times 3} - 3\sqrt{3}\\ &= 2 \times 2\sqrt{3} - 3\sqrt{3}\\ &= 4\sqrt{3} - 3\sqrt{3}\\ &= \sqrt{3}\end{aligned}$$

$$\begin{aligned}\sqrt{8} + \sqrt{50} &= \sqrt{4 \times 2} + \sqrt{25 \times 2}\\ &= 2\sqrt{2} + 5\sqrt{2}\\ &= 7\sqrt{2}\end{aligned}$$

$$\begin{aligned}\sqrt{\frac{32}{9}} &= \frac{\sqrt{32}}{\sqrt{9}}\\ &= \frac{\sqrt{16 \times 2}}{3}\\ &= \frac{4\sqrt{2}}{3}.\end{aligned}$$

You could have written

$$\sqrt{\frac{32}{9}} = \frac{\sqrt{32}}{\sqrt{9}} = \frac{\sqrt{4 \times 8}}{3} = \frac{2\sqrt{8}}{3}$$

but this is not as simple as $\frac{4\sqrt{2}}{3}$ since $\sqrt{8}$ can itself be simplified by saying that $\sqrt{8} = \sqrt{4 \times 2} = 2\sqrt{2}$. Always ensure that you simplify surd expressions as much as possible: for expressions involving square roots of a number ensure that the number inside the square root has no factors that are square numbers.

The usual rules of algebra and indices allow you to multiply expressions involving surds:

$$\sqrt{3} \times \sqrt{5} = \sqrt{3 \times 5} = \sqrt{15}$$

$$\sqrt{5}(4 - \sqrt{5}) = 4\sqrt{5} - (\sqrt{5})^2 = 4\sqrt{5} - 5$$

Remember that if a is positive then $(\sqrt{a})^2 = a$.

$$\sqrt{5}(4 - \sqrt{7}) = 4\sqrt{5} - \sqrt{5} \times \sqrt{7} = 4\sqrt{5} - \sqrt{35}$$

$$\begin{aligned}(2 + \sqrt{7})(4 - \sqrt{7}) &= 8 - 2\sqrt{7} + 4\sqrt{7} - (\sqrt{7})^2\\ &= 8 + 2\sqrt{7} - 7\\ &= 1 + 2\sqrt{7}\end{aligned}$$

$$\begin{aligned}(4 - \sqrt{5})^2 &= (4 - \sqrt{5})(4 - \sqrt{5})\\ &= 16 - 4\sqrt{5} - 4\sqrt{5} + (\sqrt{5})^2\\ &= 16 - 8\sqrt{5} + 5\\ &= 21 - 8\sqrt{5}.\end{aligned}$$

For divisions or fractions where the denominator is a simple surd, the expression can be simplified by multiplying the top and bottom of the expression by the denominator. This process is called "rationalising the denominator" since it has the effect of removing the surds from the denominator:

$$\frac{8}{\sqrt{5}} = \frac{8 \times \sqrt{5}}{\sqrt{5} \times \sqrt{5}} = \frac{8\sqrt{5}}{5}.$$

Multiplying the top and bottom of a fraction by $\sqrt{5}$ doesn't change the value of the fraction.

Similarly,

$$\frac{3\sqrt{2}}{\sqrt{6}} = \frac{3\sqrt{2} \times \sqrt{6}}{\sqrt{6} \times \sqrt{6}} = \frac{3\sqrt{12}}{6} = \frac{\sqrt{12}}{2} = \frac{\sqrt{4 \times 3}}{2} = \frac{2\sqrt{3}}{2} = \sqrt{3}.$$

These properties of surds can be used to simplify the exact answers that you have obtained when solving quadratic equations or calculating distances between points in co-ordinate geometry.

EXAMPLE 1

If A and B are the points $(-1, 2\sqrt{3})$ and $(3, 4\sqrt{3})$, respectively, prove that the distance AB is $2\sqrt{7}$.

SOLUTION

$$\begin{aligned} AB &= \sqrt{(3-(-1))^2 + (4\sqrt{3} - 2\sqrt{3})^2} \\ &= \sqrt{4^2 + (2\sqrt{3})^2} \\ &= \sqrt{16 + 12} \\ &= \sqrt{28} \\ &= \sqrt{4 \times 7} \\ &= 2\sqrt{7}. \end{aligned}$$

Remember that the distance between (a, b) and (c, d) is $\sqrt{(c-a)^2 + (d-b)^2}$.

$(2\sqrt{3})^2 = 2^2(\sqrt{3})^2 = 4 \times 3 = 12$

EXAMPLE 2

Find the exact values of the roots of the equation $t^2 - 8t - 38 = 0$.

Using the formula for quadratic equations with $a = 1$, $b = -8$, $c = -38$ gives

SOLUTION

$$\begin{aligned} t &= \frac{-b \pm \sqrt{b^2 - 4ac}}{2a} \\ &= \frac{8 \pm \sqrt{(-8)^2 - 4 \times 1 \times (-38)}}{2} \\ &= \frac{8 \pm \sqrt{64 - (-152)}}{2} \\ &= \frac{8 \pm \sqrt{216}}{2} \\ &= \frac{8 \pm 6\sqrt{6}}{2} \\ &= 4 + 3\sqrt{6} \text{ or } 4 - 3\sqrt{6}. \end{aligned}$$

$\sqrt{216} = \sqrt{36 \times 6} = 6\sqrt{6}$

EXERCISE 1

1 Simplify the following expressions as much as possible:

a) $\sqrt{75}$ **b)** $\sqrt{54}$ **c)** $\sqrt{18}$ **d)** $\sqrt{72}$ **e)** $\sqrt{3200}$ **f)** $\sqrt{\frac{18}{25}}$

g) $\sqrt{\frac{75}{121}}$ **h)** $\sqrt{\frac{64}{75}}$ **i)** $\sqrt{\frac{40}{9}}$ **j)** $\sqrt{\frac{25}{9}}$ **k)** $\sqrt{\frac{16}{3}}$ **l)** $\sqrt{240}$

2 Simplify the following expressions as much as possible:

a) $\sqrt{8}+\sqrt{18}$ **b)** $\sqrt{75}-\sqrt{27}$ **c)** $\sqrt{20}+2\sqrt{45}$ **d)** $\sqrt{18}+\sqrt{27}$ **e)** $\sqrt{28}-\sqrt{12}$

3 If $p = 3+\sqrt{5}$, $q = 7-2\sqrt{5}$ find simplified expressions for

a) $p+q$ **b)** $3p-2q$

4 Expand and simplify the following expressions:

a) $\sqrt{3}(1+4\sqrt{3})$ **b)** $\sqrt{2}(3-\sqrt{2})+2(9+\sqrt{2})$ **c)** $(2+\sqrt{5})(3-\sqrt{5})$
d) $(3\sqrt{2}-1)(5-\sqrt{2})$ **e)** $(3+\sqrt{7})(3-\sqrt{7})$ **f)** $(7+2\sqrt{3})(7-2\sqrt{3})$
g) $(1+\sqrt{7})^2$ **h)** $(9-2\sqrt{5})^2$

5 Simplify the following:

a) $\dfrac{3}{\sqrt{7}}$ **b)** $\dfrac{2}{\sqrt{10}}$ **c)** $\dfrac{6}{\sqrt{9}}$ **d)** $\dfrac{5}{\sqrt{10}}$
e) $\dfrac{1}{\sqrt{5}}$ **f)** $\dfrac{4}{\sqrt{2}}+3\sqrt{2}$ **g)** $\dfrac{21}{\sqrt{3}}-\sqrt{27}$ **h)** $\left(\dfrac{4}{\sqrt{7}}\right)^2$

6 Find the exact values of the roots of the following quadratic equations, expressing your answers as simply as possible:

a) $x^2-4x-71=0$ **b)** $y^2+6y-9=0$

7 Find the distance between the points $(3\sqrt{5}, 2\sqrt{7})$ and $(5\sqrt{5}, 0)$, giving your answer in surd form, as simply as possible.

8 By using suitable numbers for p and q, show that the following statements are not always true:

i) $\sqrt{p+q}=\sqrt{p}+\sqrt{q}$ **ii)** $\sqrt{p-q}=\sqrt{p}-\sqrt{q}$ **iii)** $\sqrt{p^2}=p$

9 Expand and simplify the following expressions:

a) $(5+2\sqrt{2})(5-2\sqrt{2})$ **b)** $(9-5\sqrt{3})(9+5\sqrt{3})$
c) $(7+2\sqrt{5})(7-2\sqrt{5})$ **d)** $(a+b\sqrt{x})(a-b\sqrt{x})$

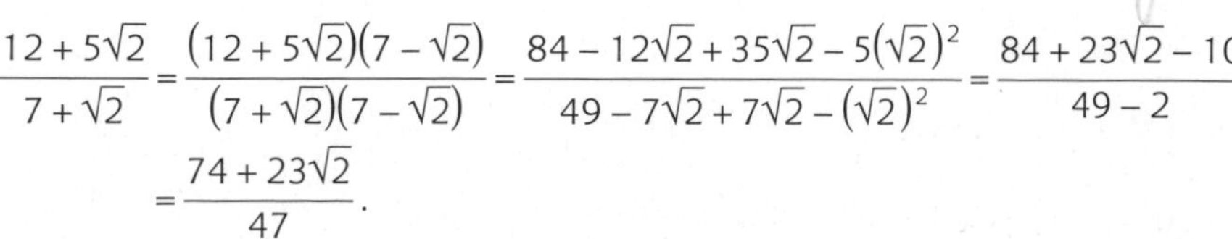

Division of an expression by $a + b\sqrt{x}$

The expression $\dfrac{12+5\sqrt{2}}{7+\sqrt{2}}$ can be simplified by multiplying the top and bottom of the fraction by $7-\sqrt{2}$:

$$\frac{12+5\sqrt{2}}{7+\sqrt{2}}=\frac{(12+5\sqrt{2})(7-\sqrt{2})}{(7+\sqrt{2})(7-\sqrt{2})}=\frac{84-12\sqrt{2}+35\sqrt{2}-5(\sqrt{2})^2}{49-7\sqrt{2}+7\sqrt{2}-(\sqrt{2})^2}=\frac{84+23\sqrt{2}-10}{49-2}$$
$$=\frac{74+23\sqrt{2}}{47}.$$

Notice how multiplying top and bottom by $7-\sqrt{2}$ produced a denominator which was an integer. (In other words, multiplying top and bottom by $7-\sqrt{2}$ had the effect of rationalising the denominator.)

In general, if a and b are integers, and x is a positive integer, then multiplying $a+b\sqrt{x}$ by $a-b\sqrt{x}$ will give an integer since

$$(a+b\sqrt{x})(a-b\sqrt{x})=a^2-ab\sqrt{x}+ab\sqrt{x}-b^2(\sqrt{x})^2$$
$$=a^2-b^2x^2.$$

This is really just an example of the "difference of two squares" result $(p+q)(p-q)\equiv p^2-q^2$.

The first step in simplifying $\dfrac{c + d\sqrt{x}}{a + b\sqrt{x}}$ is to multiply both the numerator and the denominator of the fraction by $a - b\sqrt{x}$. This will have the effect of producing a denominator which is an integer.

The expression $\dfrac{5 - \sqrt{5}}{6 + 2\sqrt{5}}$ can be simplified by multiplying the top and bottom of the fraction by $6 - 2\sqrt{5}$:

$$\frac{5-\sqrt{5}}{6+2\sqrt{5}} = \frac{(5-\sqrt{5})(6-2\sqrt{5})}{(6+2\sqrt{5})(6-2\sqrt{5})} = \frac{30-10\sqrt{5}-6\sqrt{5}+2(\sqrt{5})^2}{36-12\sqrt{5}+12\sqrt{5}-4(\sqrt{5})^2} = \frac{30-16\sqrt{5}+10}{36-20}$$

$$= \frac{40-16\sqrt{5}}{16}$$

$$= \frac{5-2\sqrt{5}}{2}.$$

Similarly, the expression $\dfrac{2+\sqrt{3}}{5-\sqrt{3}}$ can be simplified by multiplying the top and bottom of the fraction by $5+\sqrt{3}$:

$$\frac{2+\sqrt{3}}{5-\sqrt{3}} = \frac{(2+\sqrt{3})(5+\sqrt{3})}{(5-\sqrt{3})(5+\sqrt{3})} = \frac{10+2\sqrt{3}+5\sqrt{3}+(\sqrt{3})^2}{25+5\sqrt{3}-5\sqrt{3}-(\sqrt{3})^2} = \frac{10+7\sqrt{3}+3}{25-3} = \frac{13+7\sqrt{3}}{22}.$$

As well as performing arithmetic operations on expressions involving surds, it is also possible to find the exact solutions of equations involving surds.

EXAMPLE 3

Solve the equation $(4-\sqrt{3})y = 14 + 3\sqrt{3}$.

SOLUTION

$$(4-\sqrt{3})y = 14+3\sqrt{3}$$

$$\Rightarrow \quad y = \frac{14+3\sqrt{3}}{(4-\sqrt{3})}$$

$$\Rightarrow \quad y = \frac{(14+3\sqrt{3})(4+\sqrt{3})}{(4-\sqrt{3})(4+\sqrt{3})} = \frac{56+14\sqrt{3}+12\sqrt{3}+9}{16+4\sqrt{3}-4\sqrt{3}-3} = \frac{65+26\sqrt{3}}{13}$$

$$\Rightarrow \quad y = 5+2\sqrt{3}.$$

EXAMPLE 4

Solve the equation $(2+3\sqrt{2})x + 1 = (1+\sqrt{2})x + 5\sqrt{2}$.

SOLUTION

$$(2+3\sqrt{2})x + 1 = (1+\sqrt{2})x + 5\sqrt{2}$$

$$\Rightarrow \quad (2+3\sqrt{2})x - (1+\sqrt{2})x = 5\sqrt{2} - 1$$

Collect all the x terms on one side of the equation and the remaining terms on the other side.

$$\Rightarrow \quad ((2+3\sqrt{2}) - (1+\sqrt{2}))x = 5\sqrt{2} - 1$$

$$\Rightarrow \quad (1+2\sqrt{2})x = 5\sqrt{2} - 1$$

$$\Rightarrow \quad x = \frac{5\sqrt{2}-1}{(1+2\sqrt{2})}$$

EXAMPLE 4 (continued)

$$\Rightarrow \quad x = \frac{5\sqrt{2}-1}{(1+2\sqrt{2})}$$

$$\Rightarrow \quad x = \frac{(5\sqrt{2}-1)(1-2\sqrt{2})}{(1+2\sqrt{2})(1-2\sqrt{2})} = \frac{5\sqrt{2}-20-1+2\sqrt{2}}{1-2\sqrt{2}+2\sqrt{2}-8} = \frac{-21+7\sqrt{2}}{-7} = 3-\sqrt{2}$$

EXAMPLE 5

Solve the simultaneous equations

$$x - \sqrt{2}y = 0$$
$$\sqrt{2}x + 3y = 5\sqrt{10}.$$

SOLUTION

Multiplying the first equation by $\sqrt{2}$ and leaving the second equation alone will balance the coefficients of x:

$$x - \sqrt{2}y = 0 \quad \xrightarrow{\times\sqrt{2}} \quad \sqrt{2}x - 2y = 0$$
$$\sqrt{2}x + 3y = 5\sqrt{10} \quad \longrightarrow \quad \sqrt{2}x + 3y = 5\sqrt{10}$$
$$-5y = -5\sqrt{10}$$
$$y = \sqrt{10}.$$

The first equation gives

$$x = \sqrt{2}y$$
$$= \sqrt{2} \times \sqrt{10} = \sqrt{20} = \sqrt{4 \times 5}$$
$$= 2\sqrt{5}.$$

EXAMPLE 6

Find the exact values of the roots of the equation $x^2 - 8\sqrt{3}x + 36 = 0$.

SOLUTION

Using the formula for quadratic equations with $a = 1$, $b = -8\sqrt{3}$, $c = 36$ gives

$$x = \frac{-b \pm \sqrt{b^2-4ac}}{2a}$$
$$= \frac{8\sqrt{3} \pm \sqrt{(-8\sqrt{3})^2 - 4 \times 1 \times 36}}{2}$$
$$= \frac{8\sqrt{3} \pm \sqrt{192-144}}{2}$$

$(-8\sqrt{3})^2 = (-8)^2 \times (\sqrt{3})^2 = 64 \times 3 = 192$

$$= \frac{8\sqrt{3} \pm \sqrt{48}}{2}$$
$$= \frac{8\sqrt{3} \pm 4\sqrt{3}}{2}$$

$\sqrt{48} = \sqrt{16 \times 3} = 4\sqrt{3}$

$$= \frac{12\sqrt{3}}{2} \quad \text{or} \quad \frac{4\sqrt{3}}{2}$$
$$= 6\sqrt{3} \quad \text{or} \quad 2\sqrt{3}.$$

EXERCISE 2

1 **a)** Simplify $\dfrac{3}{8+\sqrt{2}}$ by multiplying the top and bottom of the fraction by $8-\sqrt{2}$.

b) Simplify $\dfrac{5+2\sqrt{7}}{10-\sqrt{7}}$ by multiplying the top and bottom of the fraction by $10+\sqrt{7}$.

c) Simplify $\dfrac{3-2\sqrt{2}}{5+2\sqrt{2}}$ by multiplying the top and bottom of the fraction by $5-2\sqrt{2}$.

2 Simplify

a) $\dfrac{9}{3+\sqrt{2}}$ **b)** $\dfrac{4+7\sqrt{5}}{10-\sqrt{5}}$ **c)** $\dfrac{7-\sqrt{3}}{2+\sqrt{3}}$ **d)** $\dfrac{8+2\sqrt{7}}{11-2\sqrt{7}}$ **e)** $\dfrac{7+2\sqrt{10}}{9+4\sqrt{10}}$

3 A rectangle has width $(7+\sqrt{3})$ cm and area $(61-11\sqrt{3})$ cm^2.
Find the length of the rectangle and prove that the perimeter of the rectangle is $(34-4\sqrt{3})$ cm.

4 Prove that

a) $\sqrt{72}+\dfrac{27+4\sqrt{2}}{5+2\sqrt{2}}-\sqrt{32}=7$ **b)** $(1+\sqrt{12})(\sqrt{75}-3)=27-\sqrt{3}$

c) $\dfrac{4+\sqrt{5}}{1+\sqrt{5}}+\dfrac{4-\sqrt{5}}{1-\sqrt{5}}=\dfrac{1}{2}$ **d)** $\dfrac{4+\sqrt{5}}{1+\sqrt{5}}+\dfrac{4-\sqrt{5}}{3-\sqrt{5}}=2+\sqrt{5}$

5 ABC is a right angled triangle with $\angle B = 90°$. The area of the triangle is 18 cm^2.
The length of AB is $(7-\sqrt{13})$ cm.

a) Prove that the length of BC is $(7+\sqrt{13})$ cm.

b) Prove that the length of AC is $2\sqrt{31}$ cm.

6 Simplify

a) $\sqrt{28}+\sqrt{63}$ **b)** $\dfrac{12}{\sqrt{10}}$ **c)** $(3+4\sqrt{2})(7-2\sqrt{2})$ **d)** $(6-2\sqrt{3})^2$ **e)** $\dfrac{2+3\sqrt{5}}{3-\sqrt{5}}$

7 **a)** Solve the equation $(4-\sqrt{5})w = 23-3\sqrt{5}$ giving your answer in the form $z=\alpha+\beta\sqrt{5}$, where α and β are integers.

b) Solve the equation $(2+\sqrt{3})x-5=4\sqrt{3}$ giving your answer in the form $x=a+b\sqrt{3}$ where a and b are integers.

c) Solve the equation $24+\sqrt{7}y=2y+9\sqrt{7}$ giving your answer in the form $y=p+q\sqrt{7}$ where p and q are integers.

8 **a)** Find the exact solutions of the simultaneous equations $\left.\begin{aligned}\sqrt{3}p+\sqrt{2}q&=10\\ \sqrt{2}p-\sqrt{3}q&=0\end{aligned}\right\}.$

b) Find the exact solutions of the simultaneous equations $\left.\begin{aligned}\sqrt{3}p+2q&=11-\sqrt{3}\\ 5p-\sqrt{3}q&=8+\sqrt{3}\end{aligned}\right\}.$

9 Find the exact values of the roots of the following equations, expressing your answers as simply as possible:

a) $z^2-4\sqrt{5}z-60=0$ **b)** $2z^2+7\sqrt{3}z-12=0$

Having studied this chapter you should know how

- to use the rules $\sqrt{pq} = \sqrt{p}\sqrt{q}$ $\sqrt{\frac{p}{q}} = \frac{\sqrt{p}}{\sqrt{q}}$

  and the basic rules of algebra to simplify and work with surd expressions
- to simplify $\frac{p}{\sqrt{s}}$ by multiplying top and bottom by $\sqrt{s}$
- to simplify $\frac{a+b\sqrt{s}}{c+d\sqrt{s}}$ by multiplying top and bottom by $c-d\sqrt{s}$

REVISION EXERCISE

1 Simplify

a) $\sqrt{72} - \frac{8}{\sqrt{2}}$ **b)** $(6-2\sqrt{5})(1+4\sqrt{5})$ **c)** $(3-\sqrt{7})^2$ **d)** $\frac{3+\sqrt{5}}{4-\sqrt{5}}$

2 a) Find the value of k if $\sqrt{72}+\sqrt{50}=k\sqrt{2}$.

b) Find the value of k if $\sqrt{48}+3\sqrt{75}-\frac{18}{\sqrt{27}}=k\sqrt{3}$.

3 A right angled triangle has area 24 cm^2 and base $8\sqrt{3}$ cm. Find the height of the triangle and prove that the hypotenuse has length $2\sqrt{51}$ cm.

4 Solve the equation $x\sqrt{8}-11=\frac{3x}{\sqrt{2}}$, giving your answer in the form $k\sqrt{2}$ where k is an integer.

(OCR May 2002 P1)

5 Find the exact solutions of the simultaneous equations $\left.\begin{array}{r}\sqrt{12}p+\sqrt{3}q=15\\ p-q=7\sqrt{3}\end{array}\right\}$.

6 Find the exact values of the roots of the equation $x^2+8x-29=0$, giving your answers in the simplest possible form.

7 a) Simplify $(5-2\sqrt{3})(4+\sqrt{3})$.

b) Express $\frac{-10+4\sqrt{7}}{4+2\sqrt{7}}$ in the form $a+b\sqrt{7}$ where a and b are integers.

c) Solve the equation $(4-\sqrt{2})x+1-3\sqrt{2}=3+14\sqrt{2}$, giving the solution in the form $x=p+q\sqrt{2}$ where p and q are integers.

8 If $\gamma=\frac{1+\sqrt{5}}{2}$ prove that

a) $\gamma^2=1+\gamma$ **b)** $\frac{1}{\gamma}=\gamma-1$

9 Solve the equation $x^2-10\sqrt{3}x+63=0$, giving your answers in terms of surds, simplified as far as possible.

10 Prove that the distance between the points $(\sqrt{5}, 2\sqrt{3})$ and $(2\sqrt{5}, -\sqrt{3})$ is $4\sqrt{2}$.

6 The Quadratic Function

The purpose of this chapter is to enable you to

- sketch graphs of quadratic functions in cases when the function can be factorised and in cases when the function cannot be factorised
- solve quadratic inequalities
- use the discriminant to determine the number of real solutions that a quadratic equation possesses

Sketching the Graph of a Quadratic Function

The diagram shows the graph of $y = 2x^2 + 7x - 9$:

Notice that the graph is ∪ shaped with a line of symmetry passing through the vertex, or minimum point, of the curve.

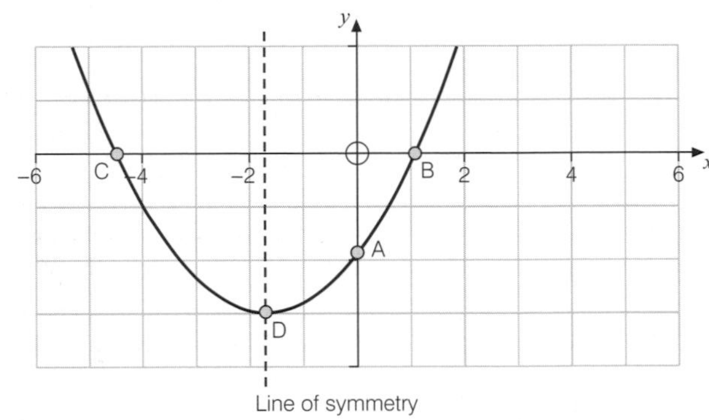

If it is possible to establish a procedure for determining:

- whether a quadratic graph is ∪ or ∩ shaped,
- where the graph crosses the co-ordinate axes,
- where the vertex (the maximum or minimum point) of the graph is,

then it would be possible to know the precise co-ordinates of the points A, B, C and D on the graph and you would be able to sketch the graph rapidly.

You can start by using a computer graph drawing package to gain an indication of the types of graph that can be obtained from quadratic functions:

$y = x^2$

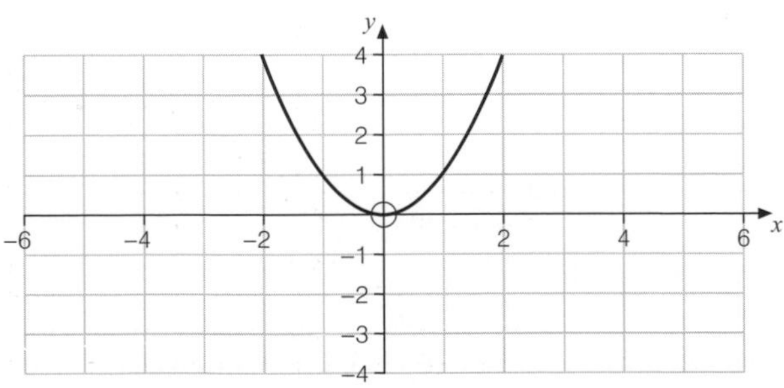

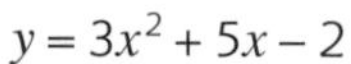

$y = 3x^2 + 5x - 2$

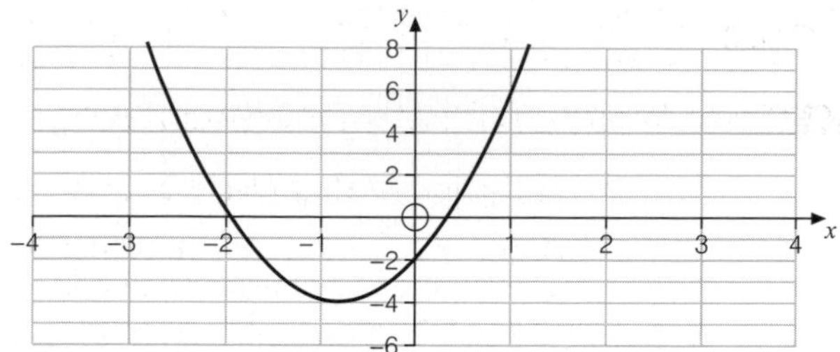

$y = 0.2x^2 + 3.2x - 1.8$

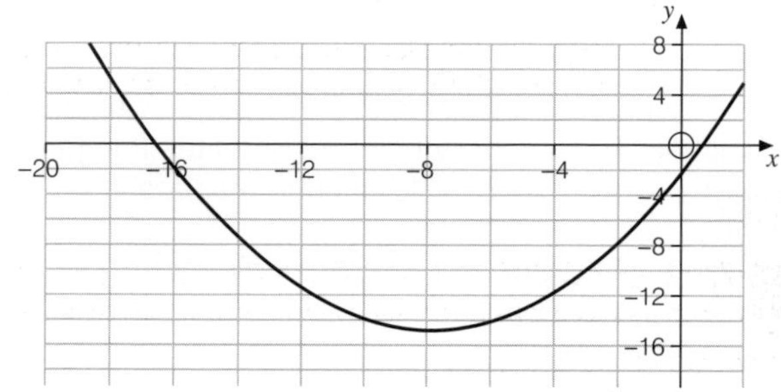

$y = -x^2$

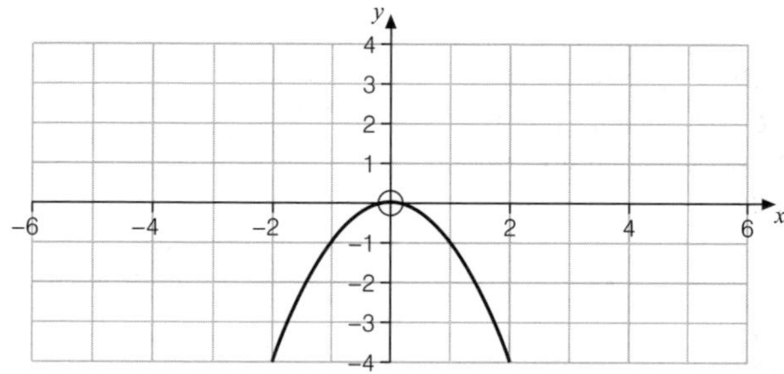

$y = -4x^2 + 7x - 10$

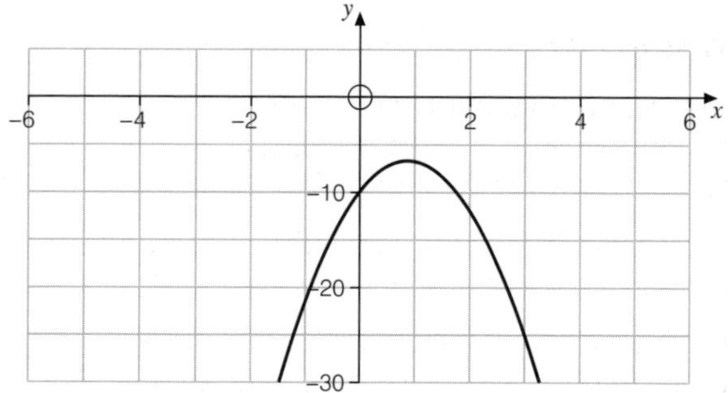

$y = -0.5x^2 + 3.2x$

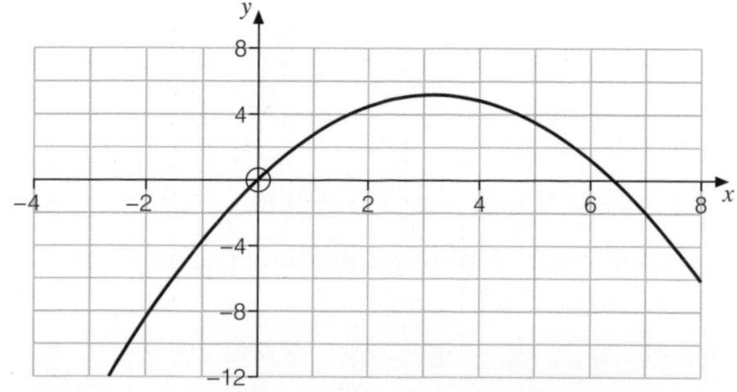

From these examples it should be clear that

> The graph of $y = ax^2 + bx + c$
> - is ∪ shaped if $a > 0$
> - is ∩ shaped if $a < 0$

Sketching the Graph of a Quadratic Function that can be Factorised

If the quadratic function is factorisable then the points where the graph meets the x axis may be calculated rapidly. The line of symmetry bisects the line joining these two points so the co-ordinates of the vertex may be calculated immediately.

EXAMPLE 1

Sketch the graph of $y = 3x^2 + 6x - 45$.

SOLUTION

- Since the coefficient of x^2 is positive, the graph will be ∪ shaped.
- When $x = 0$, $y = -45$.

Notice that $y = 3x^2 + 6x - 45 = 3(x^2 + 2x - 15) = 3(x + 5)(x - 3)$.

- The graph meets the x axis when $y = 0$

$$\Rightarrow \quad 3(x+5)(x-3) = 0$$
$$\Rightarrow \quad x+5=0 \quad \text{or} \quad x-3=0$$
$$\Rightarrow \quad x=-5 \quad \text{or} \quad x=3.$$

- The line of symmetry for the graph is $x = -1$.

$\dfrac{-5+3}{2} = -1$

When $x = -1$, $y = 3(-1)^2 + 6(-1) - 45 = 3 + (-6) - 45 = -48$
so the minimum point is $(-1, -48)$
and it is now possible to sketch the graph:

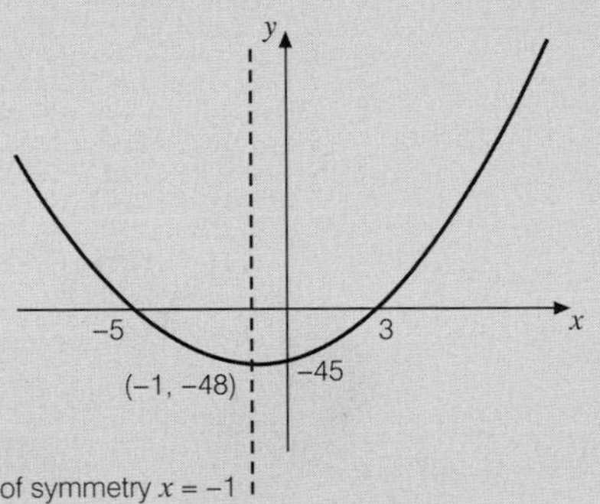

EXAMPLE 2

Factorise $-5x^2 + 8x + 4$ and hence sketch the graph of $y = -5x^2 + 8x + 4$.

SOLUTION

The negative coefficient of x^2 requires some care: one way of handling it is to factor it out at the beginning:

$$\begin{aligned} -5x^2 + 8x + 4 &= -(5x^2 - 8x - 4) \\ &= -(5x+2)(x-2). \end{aligned}$$

EXAMPLE 2 (continued)

- Since the coefficient of x^2 is negative, the graph will be ∩ shaped.
- When $x = 0$, $y = 4$.

Notice that $y = -5x^2 + 8x + 4 = -(5x + 2)(x - 2)$.

- The graph meets the x axis when $y = 0$

$$\Rightarrow \quad -(5x+2)(x-2) = 0$$
$$\Rightarrow \quad 5x+2 = 0 \quad \text{or} \quad x-2 = 0$$
$$\Rightarrow \quad x = -\tfrac{2}{5} \quad \text{or} \quad x = 2.$$

$$\frac{-\frac{2}{5}+2}{2} = \frac{4}{5}$$

- The line of symmetry for the graph is $x = \frac{4}{5}$. When $x = \frac{4}{5}$, $y = -(5 \times \frac{4}{5} + 2)(\frac{4}{5} - 2) = -6 \times -\frac{6}{5} = +\frac{36}{5}$ so the maximum point of the curve for example 2 is the point $(\frac{4}{5}, +\frac{36}{5})$ and we can now sketch the graph:

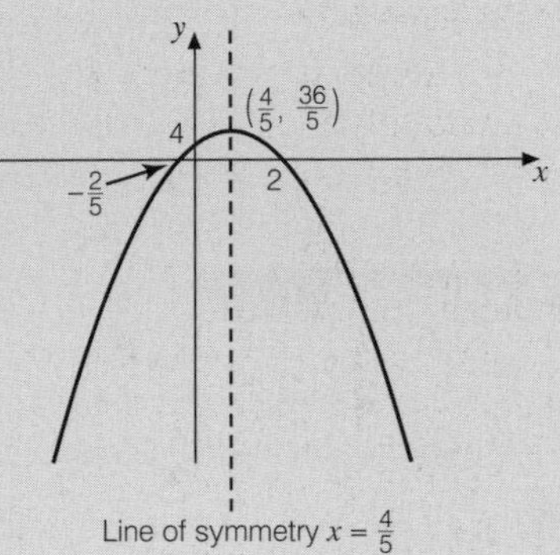

EXAMPLE 3

The diagram shows the graph of $y = k(x - a)(x - b)$ where k, a and b are constants with $a < b$.

a) Determine the values of the constants a, b and k.
b) Determine the co-ordinates of the vertex of the graph.

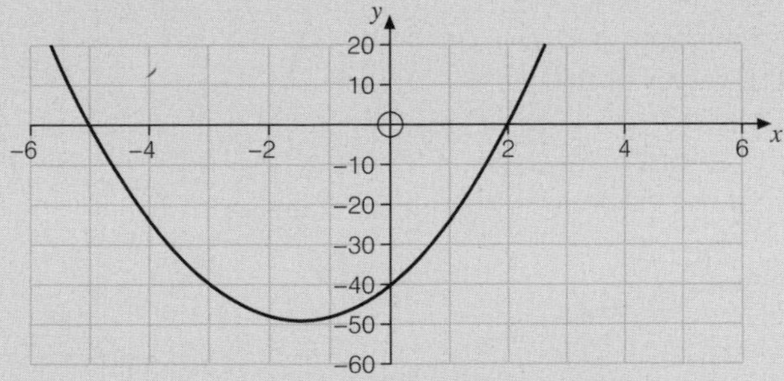

SOLUTION

a) The graph of $y = k(x - a)(x - b)$ crosses the x axis when

$$k(x-a)(x-b) = 0$$
$$\Rightarrow \quad x-a = 0 \quad \text{or} \quad x-b = 0$$
$$\Rightarrow \quad x = a \quad \text{or} \quad x = b.$$

From the diagram, you can see that the curve crosses the x axis at −5 and 2.

Since $a < b$, then $a = -5$ and $b = 2$.

EXAMPLE 3 (continued)

The graph of $y = k(x - a)(x - b)$ crosses the y axis at the point $(0, kab)$.
From the diagram, you can see that the graph crosses the y axis at $(0, -40)$.

Therefore

$$kab = -40$$
$$\Rightarrow \quad k \times (-5) \times 2 = -40$$
$$\Rightarrow \quad -10k = -40$$
$$\Rightarrow \quad k = 4.$$

b) The vertex lies on the line of symmetry which is $x = \frac{-5+2}{2} = -\frac{3}{2}$.
When $x = -\frac{3}{2}$, $y = 4 \times ((-\frac{3}{2}) - (-5))((-\frac{3}{2}) - 2) = 4 \times \frac{7}{2} \times (-\frac{7}{2}) = -49$ so the vertex is the point $(-\frac{3}{2}, -49)$.

EXERCISE 1

In questions 1–6 sketch the graphs, taking care to show where the graph meets the axes and the co-ordinates of the vertex.

1 $y = (x - 1)(x - 5)$ **2** $y = (x + 3)(x - 5)$ **3** $y = (4 - x)(x + 6)$

4 $y = 3(x - 2)(x + 4)$ **5** $y = (2x - 1)(x + 4)$ **6** $y = 5(6 - x)(x - 2)$

In questions 7–12, write the equations of the curves in factorised form and hence sketch the graphs.

7 $y = x^2 + 5x - 50$ **8** $y = 2x^2 + 8x - 42$ **9** $y = 24 + 2x - x^2$

10 $y = 3x^2 + 7x - 26$ **11** $y = 90 + 15x - 5x^2$ **12** $y = 10 + 3x - 4x^2$

13 The diagram shows a sketch of $y = k(x - p)(x - q)$ where k, p and q are constants with $p < q$.

a) Determine the values of the constants p, q and k.

b) Determine the co-ordinates of the vertex of the graph.

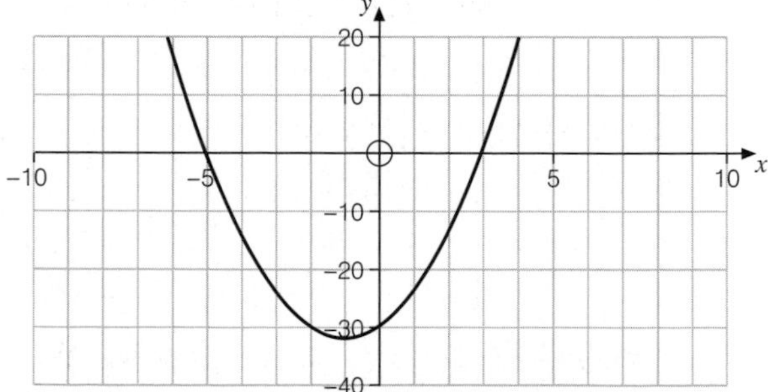

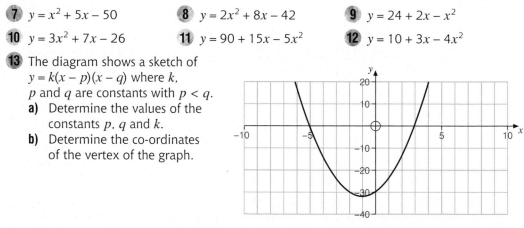

14 The diagram shows a sketch of $y = k(x - c)(x - d)$ where k, c and d are constants with $c < d$. The point (2, 18) is the maximum point of the graph.

a) Determine the values of the constants c, d and k.

b) Determine the co-ordinates of the point where the graph crosses the y axis.

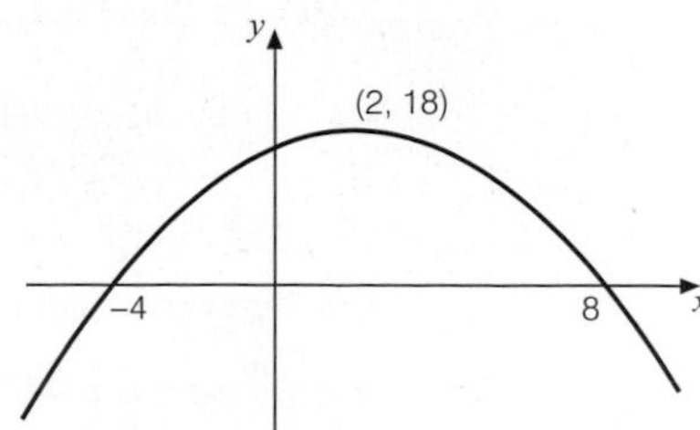

 15 The diagram shows a sketch of $y = k(x - s)(x - t)$ where k, s and t are constants with $s < t$. The point (2.5, −11.25) is the minimum point of the graph.

a) Determine the values of the constants s, t and k.

b) Determine the co-ordinates of the point where the graph crosses the y axis.

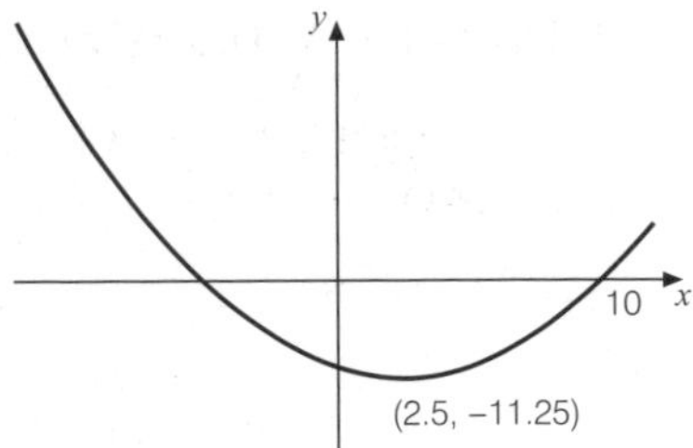

The Completed Square Format for $ax^2 + bx + c$

You have already seen in chapter 3 how expressions of the form $x^2 + bx + c$ can be rewritten in the form $(x + q)^2 + r$ and that this was very useful in solving quadratic equations that could not be factorised.

In order to sketch the graphs of quadratic functions that cannot be factorised it is necessary to obtain the completed square format for $ax^2 + bx + c$, which is an expression of the form $p(x + q)^2 + r$.

The values of p, q and r can be found in three different ways. Each method requires care and attention to avoid careless arithmetic or algebraic mistakes. You must be able to confidently use one of these three methods.

The three methods are illustrated in the following example.

EXAMPLE 4

Write the expression $5x^2 - 12x + 4$ in the form $p(x + q)^2 + r$.

SOLUTION

Method 1: Factor out the 5 from the x^2 and x terms first, then complete the square and then expand to get final form.

$$5x^2 - 12x + 4 \equiv 5[x^2 - 2.4x] + 4$$

Completing the square gives $x^2 - 2.4x \equiv (x - 1.2)^2 - 1.44$

$$\Rightarrow \quad 5x^2 - 12x + 4 \equiv 5[(x - 1.2)^2 - 1.44] + 4$$

$$\Rightarrow \quad 5x^2 - 12x + 4 \equiv 5(x - 1.2)^2 - 7.2 + 4$$

Take care: It is easy to forget to multiply the 1.44 by 5 at this stage!

$$\Rightarrow \quad 5x^2 - 12x + 4 \equiv 5(x - 1.2)^2 - 3.2.$$

Method 2: Let y be the given expression; divide through by 5 to return to the simple case of completing the square; complete the square; then multiply by 5 to return to the original expression.

Let $\quad y \equiv 5x^2 - 12x + 4$

$[\div 5] \quad \Rightarrow \quad \dfrac{y}{5} \equiv x^2 - 2.4x + 0.8$

EXAMPLE 4 (continued)

Completing the square gives $x^2 - 2.4x \equiv (x - 1.2)^2 - 1.44$

$$\Rightarrow \quad \frac{y}{5} \equiv (x - 1.2)^2 - 1.44 + 0.8$$

$$\Rightarrow \quad \frac{y}{5} \equiv (x - 1.2)^2 - 0.64$$

$$[\times 5] \quad \Rightarrow \quad y \equiv 5(x - 1.2)^2 - 3.2$$

$$\Rightarrow \quad 5x^2 - 12x + 4 \equiv 5(x - 1.2)^2 - 3.2.$$

Method 3: Expand $p(x + q)^2 + r$ and then compare coefficients.

$$p(x+q)^2 + r \equiv p(x^2 + 2qx + q^2) + r \equiv px^2 + 2pqx + (pq^2 + r)$$

You want $5x^2 - 12x + 4$ to be the same as $p(x + q)^2 + r$.

If $5x^2 - 12x + 4 \equiv px^2 + 2pqx + (pq^2 + r)$

looking at the x^2 coefficient $\Rightarrow$ $5 = p$

looking at the x coefficient $\Rightarrow$ $-12 = 2pq$
$\Rightarrow$ $-12 = 10q$
$\Rightarrow$ $q = -1.2$

looking at the constant term $\Rightarrow$ $4 = pq^2 + r$
$\Rightarrow$ $4 = 5 \times 1.2^2 + r$
$\Rightarrow$ $4 = 7.2 + r$
$\Rightarrow$ $r = -3.2$

so $5x^2 - 12x + 4 \equiv 5(x - 1.2)^2 - 3.2$.

EXAMPLE 5

Find values p, q and r so that $-3x^2 + 9x + 2 \equiv p(x + q)^2 + r$

SOLUTION

Using method 2:

let $\quad y = -3x^2 + 9x + 2$

$$\Rightarrow \quad \frac{y}{-3} = x^2 - 3x - \frac{2}{3}$$

The completed square format of $x^2 - 3x$ is $(x - \frac{3}{2})^2 - \frac{9}{4}$.

$$\Rightarrow \quad \frac{y}{-3} = \left(x - \frac{3}{2}\right)^2 - \frac{9}{4} - \frac{2}{3}$$

$$\Rightarrow \quad \frac{y}{-3} = \left(x - \frac{3}{2}\right)^2 - \frac{35}{12}$$

$$\Rightarrow \quad y = -3\left(x - \frac{3}{2}\right)^2 + \frac{35}{4}.$$

so

$$-3x^2 + 9x + 2 \equiv -3(x - \tfrac{3}{2})^2 + \tfrac{35}{4}.$$

EXERCISE 2

Write the following expressions in completed square form $p(x + q)^2 + r$:

1 $3x^2 + 15x$	**2** $4x^2 + 24x + 2$	**3** $3x^2 - 12x + 15$
4 $6x^2 - 30x - 9$	**5** $2x^2 + 5x + 1$	**6** $4x^2 - 6x - 6$
7 $4x^2 + 9x + 3$	**8** $10x^2 - 6x - 7$	**9** $-2x^2 + 6x + 7$
10 $-5x^2 + 9x - 2$	**11** $-3x^2 + 9x - 2$	**12** $5x^2 - 8x + 7$

Sketching the Graph of a Quadratic Function that cannot be Factorised

If the curve equation cannot be factorised then you can use the method of completing the square to gain enough information to sketch the curve. Four steps will be needed.

- The graph of $y = ax^2 + bx + c$ will be ∪ shaped if $a > 0$ and ∩ shaped if $a < 0$.
- Calculate the value of y when $x = 0$.
- Write the curve equation in completed square format and hence deduce the co-ordinates of the maximum or minimum point of the curve.
- Find the points (if any) where the curve crosses the x axis.

EXAMPLE 6

Sketch the graph of $y = x^2 + 8x - 11$, taking care to show the places where the graph crosses the axes and the minimum point.

SOLUTION

Since the coefficient of x^2 is 1 which is positive, the graph will be ∪ shaped.

When $x = 0$; $y = -11$.

The method of completing the square gives

$$y = x^2 + 8x - 11 = (x + 4)^2 - 16 - 11$$

$$\Rightarrow \quad y = (x + 4)^2 - 27.$$

The minimum point on the curve is therefore the point $(-4, -27)$.

> $y = (x + 4)^2 - 27$
>
> The square expression $(x + 4)^2$ is always greater than or equal to 0 and can only be equal to 0 when $x = -4$.
>
> The smallest possible value of y is therefore
>
> $0^2 - 27 = -27$
>
> and this value is obtained when $x = -4$.
>
> The minimum point on the curve is therefore the point $(-4, -27)$.

The curve crosses the x axis where $y = 0$, which gives

$$(x + 4)^2 - 27 = 0$$

$$\Rightarrow \quad (x + 4)^2 = 27$$

$$\Rightarrow \quad x + 4 = \sqrt{27} \quad \text{or} \quad -\sqrt{27}$$

$$\Rightarrow \quad x = -4 + \sqrt{27} \quad \text{or} \quad -4 - \sqrt{27}.$$

The sketch graph of the function can now be produced:

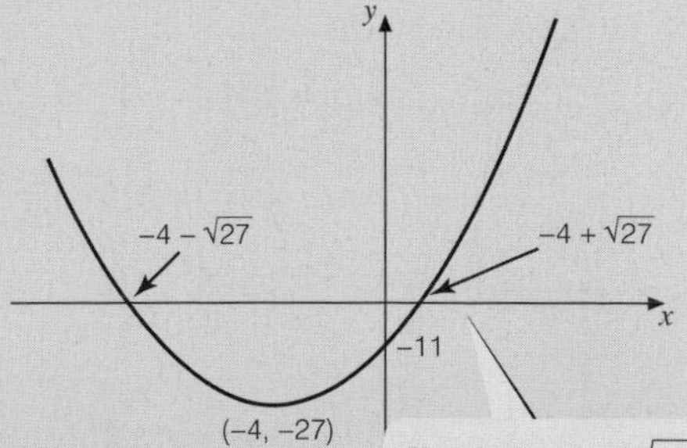

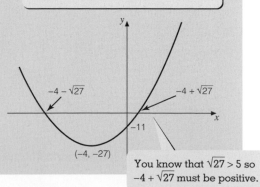

> You know that $\sqrt{27} > 5$ so $-4 + \sqrt{27}$ must be positive.

EXAMPLE 7

Sketch the graph of $y = -2x^2 + 6x + 18$, taking care to show the places where the graph crosses the axes and the maximum point.

SOLUTION

Since the coefficient of x^2 is –2 which is negative, the graph will be ∩ shaped.

When $x = 0$, $y = 18$.

The method of completing the square gives

$$y = -2x^2 + 6x + 18$$
$$\Rightarrow \quad \frac{y}{-2} = x^2 - 3x - 9$$
$$\Rightarrow \quad \frac{y}{-2} = (x - 1.5)^2 - 2.25 - 9$$
$$\Rightarrow \quad \frac{y}{-2} = (x - 1.5)^2 - 11.25$$
$$\Rightarrow \quad y = -2(x - 1.5)^2 + 22.5.$$

The curve therefore has a maximum point of (1.5, 22.5).

$$y = -2(x - 1.5)^2 + 22.5$$

Now the square expression $(x - 1.5)^2$ is always greater than or equal to 0 and can only be equal to 0 when $x = 1.5$. The expression $-2(x - 1.5)^2$ is therefore always less than or equal to 0 and can only be equal to 0 when $x = 1.5$.

So, the largest possible value of y is $0^2 + 22.5 = 22.5$ and this value is obtained when $x = 1.5$.

The maximum point on the curve is therefore the point (1.5, 22.5).

The curve crosses the x axis where $y = 0$, which gives

$$-2(x - 1.5)^2 + 22.5 = 0$$
$$\Rightarrow \quad (x - 1.5)^2 = 11.25$$
$$\Rightarrow \quad x - 1.5 = \sqrt{11.25} \quad \text{or} \quad -\sqrt{11.25}$$
$$\Rightarrow \quad x = 1.5 + \sqrt{11.25} \quad \text{or} \quad 1.5 - \sqrt{11.25}.$$

The sketch graph of the function can now be produced:

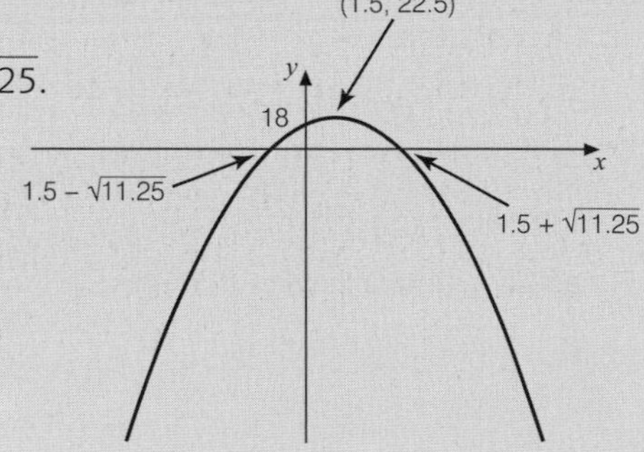

EXAMPLE 8

Sketch the graph of $y = -x^2 + 8x - 20$.

SOLUTION

Since the coefficient of x^2 is –1 which is negative, the graph will be ∩ shaped.

When $x = 0$, $y = -20$.

$$y = -x^2 + 8x - 20$$
$$\Rightarrow \quad \frac{y}{-1} = x^2 - 8x + 20$$
$$\Rightarrow \quad \frac{y}{-1} = (x - 4)^2 - 16 + 20$$
$$\Rightarrow \quad \frac{y}{-1} = (x - 4)^2 + 4$$
$$\Rightarrow \quad y = -(x - 4)^2 - 4.$$

The curve therefore has a **maximum point** (4, –4).

EXAMPLE 8 (continued)

Since the curve has a maximum of −4, it cannot cross the x axis so there are no x intercepts to be found.

The sketch graph can now be drawn:

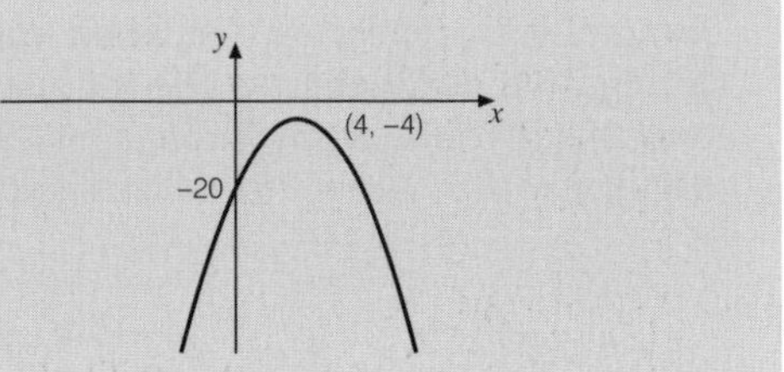

EXERCISE 3

1 Write each of the following curves in the form $y = (x + q)^2 + r$ and hence sketch its graph:
a) $y = x^2 - 4x - 7$ **b)** $y = x^2 - 12x + 47$

2 Write each of the following curves in the form $y = p(x + q)^2 + r$ and hence sketch its graph:
a) $y = 3x^3 - 24x + 5$ **b)** $y = 5x^2 - 40x + 14$
c) $y = -3x^2 - 18x + 20$ **d)** $y = -0.5x^2 + 6x + 1$

3 Write $3x^2 - 6x + 12$ in the form $p(x + q)^2 + r$ and hence write down
a) the equation of the line of symmetry of the graph $y = 3x^2 - 6x + 12$;
b) the co-ordinates of the vertex of the graph $y = 3x^2 - 6x + 12$.

4 Write $5x^2 + 8x - 4$ in the form $p(x + q)^2 + r$.
Hence write down the equation of the line of symmetry of the graph $y = 5x^2 + 8x - 4$.
Sketch the graph of $y = 5x^2 + 8x - 4$.

5 The diagram shows a sketch of the graph $y = (x + q)^2 + r$ where q and r are constants.
The minimum point on the curve is (20, −25)
a) Find the values of q and r.
b) Find the co-ordinates of the points A, B and C.

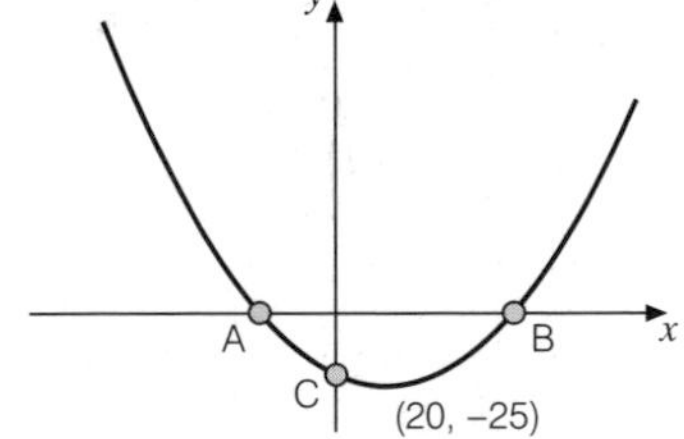

Linear and Quadratic Inequalities

Linear Inequalities

You will probably have already solved linear inequalities such as

$$5x - 2 \leqslant 0 \quad \text{or} \quad 3x - 7 \leqslant 5(x - 3) + 4.$$

The solution of such inequalities relies on the observations that if $a < b$ then

1 $a + k < b + k$ for **any** number k
2 $a - k < b - k$ for **any** number k
3 $ka < kb$ for **any positive** number k
4 $\frac{a}{k} < \frac{b}{k}$ for **any positive** number k
5 $na > nb$ for **any negative** number n
6 $\frac{a}{n} > \frac{b}{n}$ for **any negative** number n

The first four of these observations are straightforward to use but the last two need care: it is important to remember that **when you multiply or divide both sides of an inequality by a negative number then the inequality reverses**. (For example, you know that $16 < 20$ and that if you multiply through by -2 then the inequality should become $-32 > -40$. You also know that $-20 < 15$ and that if you divide through by -5 then the inequality becomes $4 > -3$.)

EXAMPLE 9

Solve the inequality $3x - 7 \leqslant 5(x - 3) + 4$.

SOLUTION

		$3x - 7 \leqslant 5(x - 3) + 4$
	$\Rightarrow$	$3x - 7 \leqslant 5x - 15 + 4$
	$\Rightarrow$	$3x - 7 \leqslant 5x - 11$
$[-3x]$	$\Rightarrow$	$-7 \leqslant 2x - 11$
$[+11]$	$\Rightarrow$	$4 \leqslant 2x$
$[\div 2]$	$\Rightarrow$	$2 \leqslant x.$

Alternatively, this inequality could be solved graphically.

$$3x - 7 \leqslant 5(x - 3) + 4 \quad \Rightarrow \quad 3x - 7 \leqslant 5x - 11.$$

The diagram shows the graphs of $y = 3x - 7$ and $y = 5x - 11$

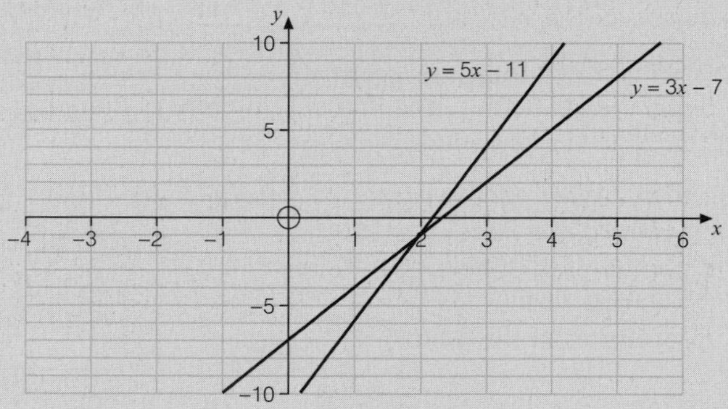

From the graph, you can see that the lines intersect at $x = 2$ and the value of $5x - 11$ is bigger than the value of $3x - 7$ when x is greater than 2. You can conclude that

$$3x - 7 \leqslant 5x - 11 \quad \Rightarrow \quad 2 \leqslant x.$$

EXAMPLE 10

Solve the inequality $17 - 2x \leqslant 5$.

SOL

		$17 - 2x \leqslant 5$
$[-17]$	$\Rightarrow$	$-2x \leqslant -12$
$[\div(-2)]$	$\Rightarrow$	$x \geqslant 6.$

Remember: dividing by -2 reverses the direction of the inequality.

Quadratic Inequalities

The graphical approach is particularly useful for solving inequalities of the form

$$f(x) < 0 \quad \text{or} \quad f(x) > 0$$

where $f(x)$ is a quadratic polynomial.

EXAMPLE 11

Solve the inequality $x^2 - 3x - 10 < 0$.

SOLUTION

The graph of $y = x^2 - 3x - 10 = (x - 5)(x + 2)$

- is ∪ shaped
- crosses the x axis at (5, 0) and (−2, 0).

Note that we do **not** need to know the co-ordinates of the vertex of the quadratic if we are just trying to solve an inequality.

We are looking for the places where the value of $x^2 - 3x - 10$ is negative: these are the points on the graph that lie **below** the x axis.

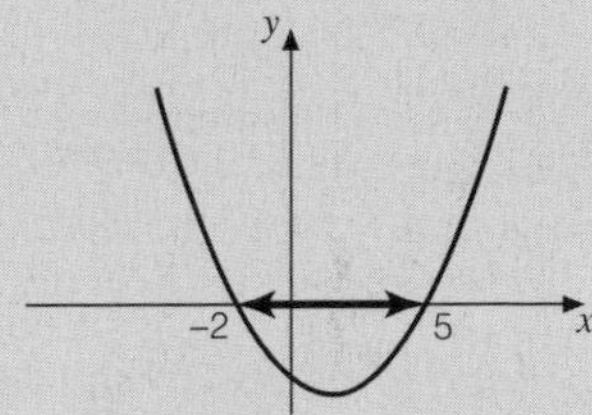

From the graph we can see that if x lies between −2 and 5 then the graph is below the x axis. In other words, the value of $x^2 - 3x - 10$ is negative if x lies between −2 and 5.

The solution to the inequality

$$x^2 - 3x - 10 < 0$$

is therefore $-2 < x < 5$.

In this example, the solution can be written as a **single** statement. The values of x that satisfy the inequality lie between −2 and 5 and this can be written as $-2 < x < 5$.

EXAMPLE 12

Solve the inequality $2x^2 \geqslant 22 - 7x$.

SOLUTION

You must start by gathering all the terms on one side of the inequality:

$$2x^2 \geqslant 22 - 7x$$

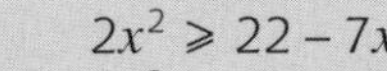

$$\Rightarrow \quad 2x^2 + 7x \geqslant 22$$

$$\Rightarrow \quad 2x^2 + 7x - 22 \geqslant 0.$$

The graph of $y = 2x^2 + 7x - 22 = (2x + 11)(x - 2)$

- is ∪ shaped
- crosses the x axis at (2, 0) and $(-\frac{11}{2}, 0)$

so a quick sketch of $y = 2x^2 + 7x - 22 = (2x + 11)(x - 2)$ can be drawn:

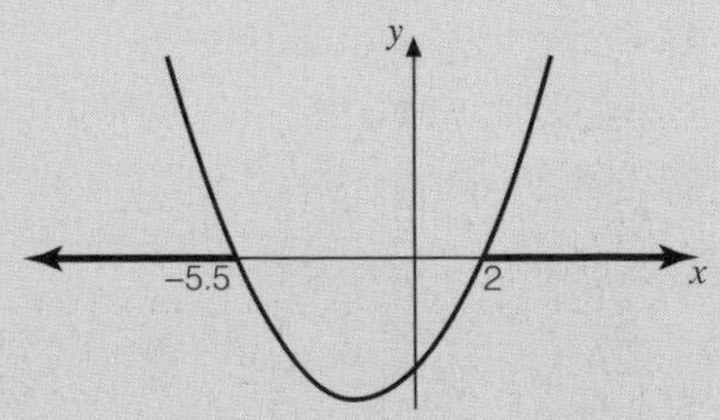

You are looking for the places where the value of $2x^2 + 7x - 22$ is greater than or equal to zero: these are the points on the graph that lie **on or above** the x axis.

From the graph you can see that the value of $2x^2 + 7x - 22$ is at least 0 if $x \geqslant 2$ or if $x \leqslant -5.5$.

$2x^2 + 7x - 22 \geqslant 0$ is therefore

$$x \geqslant 2 \quad \text{or} \quad x \leqslant -5.5.$$

Note that in this example the solution set comes in two distinct parts: **either** x is greater than or equal to 2 **or** x is less than or equal to −5.5. The final answer must therefore be written as two separate inequalities.

EXAMPLE 13

Solve the inequality $5x^2 \leqslant 3x + 3$.

SOLUTION

You need to start by gathering all the terms on one side of the inequality:

$$5x^2 \leqslant 3x + 3$$
$$\Rightarrow \quad 5x^2 - 3x - 3 \leqslant 0.$$

You will need a sketch of $y = 5x^2 - 3x - 3$:

- the graph is ∪ shaped
- when $y = 0$, $\quad 5x^2 - 3x - 3 = 0 \quad \Rightarrow \quad x = \dfrac{3 \pm \sqrt{69}}{10}$

The quadratic expression does not factorise so the formula for quadratic equations is used.

so the graph crosses the x axis at $\left(\dfrac{3+\sqrt{69}}{10}, 0\right)$ and $\left(\dfrac{3-\sqrt{69}}{10}, 0\right)$.

You are looking for the places where the value of $5x^2 - 3x - 3$ is negative: these are the points on the graph that lie **on or below** the x axis.

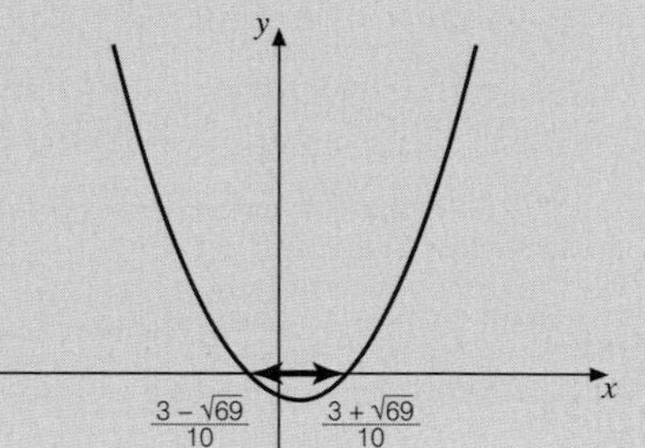

From the graph, it can be seen that

$5x^2 - 3x - 3 \leqslant 0$ when $\dfrac{3-\sqrt{69}}{10} \leqslant x \leqslant \dfrac{3+\sqrt{69}}{10}$.

EXERCISE 4

1 Solve the inequalities
a) $6x + 4 < 46$ **b)** $15 - 2x < 29$ **c)** $17 + 5x > 38 - 2x$

2 Solve the inequalities
a) $10 \leqslant 5x < 40$ **b)** $-14 < 3x - 2 < 16$

3 **a)** Find the point of intersection of the lines $y = 3x - 5$ and $y = 9 - \frac{1}{2}x$.
b) Sketch on a single diagram the lines $y = 3x - 5$ and $y = 9 - \frac{1}{2}x$.
c) Hence write down the solution of the inequality $3x - 5 < 9 - \frac{1}{2}x$.

4 Solve the inequalities
a) $x^2 + 8x - 48 < 0$ **b)** $y^2 - 8y < 20$ **c)** $3y^2 < 5y + 2$

5 Sketch the graph of $y = x^2 + 4x - 7$ and hence solve the inequality $0 < x^2 + 4x - 7$.

6 Solve the inequalities
a) $x^2 - 6x \geqslant 0$ **b)** $16 + 6x - x^2 \geqslant 0$ **c)** $(2y + 3)(y - 2) < 7y + 4$

7 Solve the inequalities
a) $(x + 5)^2 \geqslant 41 + 4x$ **b)** $(2x - 5)^2 < (x - 1)^2$

8 **a)** Solve the inequality $2x^2 + 7x + 5 \leqslant 0$.
b) Explain why the inequality $2x^2 + 7x + 7 \leqslant 0$ has no solutions.

9 A rectangular plot of land has width w m and perimeter 32 m. The area of the plot is greater than 48 m^2. Obtain and solve an inequality that must be satisfied by w.

10 A piece of string of length 18π cm is cut into two pieces. The first piece is made into a circle of radius r cm. The second piece is also made into a circle. The total area of the two circles is greater than 53π cm^2.

a) Prove that the second circle has radius $(9 - r)$ cm.

b) Prove that r satisfies $r^2 - 9r + 14 > 0$ and hence find the possible values of r.

11 The triangle ABC has $\angle B = 90°$, AB $= x$ cm and BC $= (x + 2)$ cm. The length of AC is greater than 10 cm.
Obtain and solve an inequality that must be satisfied by x.

The Discriminant of the Equation $ax^2 + bx + c = 0$

Consider the quadratic equation $x^2 - 3x + 5 = 0$.

The equation does not factorise so you might attempt to use the formula for quadratic equations:

the equation $ax^2 + bx + c = 0$ has roots given by $x = \dfrac{-b \pm \sqrt{b^2 - 4ac}}{2a}$

so the roots of $x^2 - 3x + 5 = 0$ are given by $x = \dfrac{-(-3) \pm \sqrt{(-3)^2 - 4 \times 1 \times 5}}{2 \times 1} = \dfrac{3 \pm \sqrt{-11}}{2}$.

It is not possible to calculate the square root of -11 so you cannot find any real number roots of the equation.

This can be explained by considering the graph of $y = x^2 - 3x + 5$:

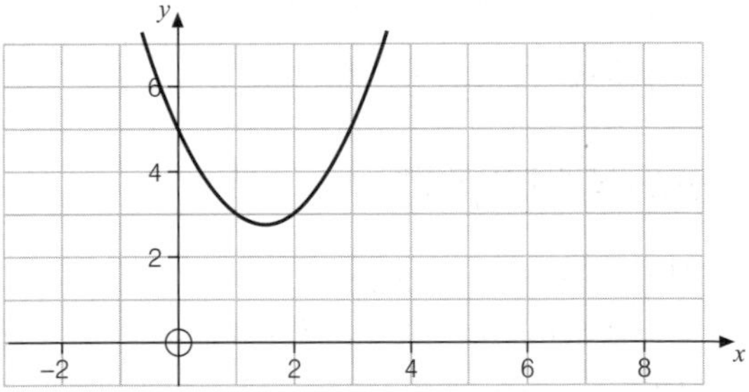

Since the minimum value of $x^2 - 3x + 5$ is greater than 0, it follows that the equation $x^2 - 3x + 5 = 0$ has no real roots.

In general, the roots of the quadratic equation $ax^2 + bx + c = 0$ are given by

$$x = \frac{-b \pm \sqrt{b^2 - 4ac}}{2a}$$

and the number of real roots depends on the value of $b^2 - 4ac$.

- If $b^2 - 4ac > 0$ then $\sqrt{b^2 - 4ac}$ is a positive number and the equation has two distinct real roots.
- If $b^2 - 4ac = 0$ then $\sqrt{b^2 - 4ac} = 0$ and the equation has just one solution $x = -\dfrac{b}{2a}$.
- If $b^2 - 4ac < 0$ then $\sqrt{b^2 - 4ac}$ cannot be evaluated since we cannot find the square root of a negative number. The equation has no real roots.

If $b^2 - 4ac$ is 0 then it is sometimes said that the equation has a **repeated** root of $-\dfrac{b}{2a}$ since the quadratic equation formula gives $x = \dfrac{-b}{2a}$ or $\dfrac{-b}{2a}$.

The number $b^2 - 4ac$ is called the **discriminant** of the equation $ax^2 + bx + c = 0$ and is often denoted by the symbol Δ.

To summarise:

For the quadratic equation $ax^2 + bx + c = 0$, the number $\Delta = b^2 - 4ac$ is called the discriminant.

If $b^2 - 4ac > 0$ then the equation has two distinct real roots;

If $b^2 - 4ac = 0$ then the equation has just one, repeated root, which is $x = -\frac{b}{2a}$;

If $b^2 - 4ac < 0$ then the equation has no real roots.

EXAMPLE 14

How many real roots does the equation $3x^2 + 7x + 5 = 0$ have?

SOL

Putting $a = 3$, $b = 7$ and $c = 5$ gives

$$\Delta = b^2 - 4ac = 49 - 4 \times 3 \times 5 = -11$$

so the equation has no real roots.

EXAMPLE 15

For what values of k does the equation $3x^2 + 7x + k = 0$ have real roots?

SOL

For real solutions $b^2 - 4ac \geqslant 0$

$$\begin{aligned} &\Rightarrow \quad 7^2 - 4 \times 3 \times k \geqslant 0 \\ &\Rightarrow \quad 49 - 12k \geqslant 0 \\ &\Rightarrow \quad 49 \geqslant 12k \\ &\Rightarrow \quad \tfrac{49}{12} \geqslant k. \end{aligned}$$

EXAMPLE 16

For what values of k does the equation $(k - 3)x^2 - kx + k = 0$ have no real roots?

SOLUTION

If the equation has no real roots then $b^2 - 4ac < 0$

$$\begin{aligned} &\Rightarrow \quad (-k)^2 - 4(k - 3)k < 0 \\ &\Rightarrow \quad k^2 - 4k^2 + 12k < 0 \\ &\Rightarrow \quad 12k - 3k^2 < 0. \end{aligned}$$

It is possible to solve this inequality in the same way as you solved the quadratic inequalities of the previous section: by sketching the graph of $y = 12k - 3k^2$ and using this to write down the solutions of $12k - 3k^2 < 0$.

The graph of $y = 12k - 3k^2 = 3k(4 - k)$

- is ∩ shaped
- $y = 0$ when $k = 0$ or $k = 4$.

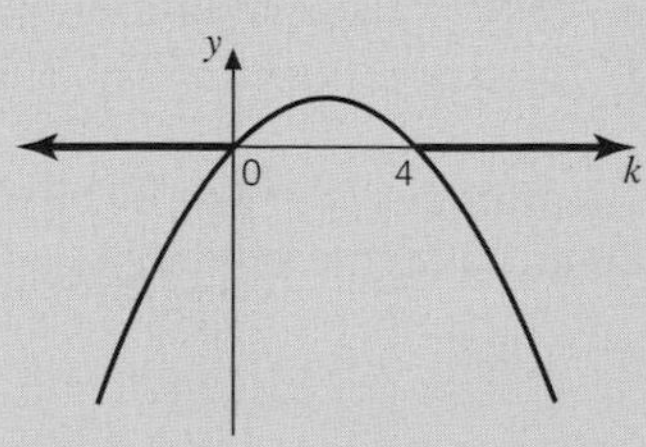

From the graph you can see that

$$\begin{aligned} &12k - 3k^2 < 0 \\ \Rightarrow \quad &k < 0 \quad \text{or} \quad k > 4. \end{aligned}$$

So the equation $(k - 3)x^2 - kx + k = 0$ has no real roots if $k < 0$ or $k > 4$.

EXERCISE 5

1 How many real roots do each of the following quadratic equations have?

a) $5x^2 - 8x + 3 = 0$ **b)** $4x^2 + 8x + 4 = 0$

c) $5x^2 + 7x + 12 = 0$ **d)** $5x^2 + 7x - 12 = 0$

2 For what values of k does the equation $2x^2 - 7x + k = 0$ have two different real roots?

3 For what values of k does the equation $3x^2 + kx + 12 = 0$ have two different real roots?

4 For what values of k does the equation $4x^2 - kx + 36 = 0$ have no real roots?

5 For what values of p does the equation $(4p + 1)x^2 - 3px + 1 = 0$ have two different real roots?

6 For what values of p does the equation $(6p + 1)x^2 - 5px + p = 0$ have two different real roots?

Having studied this chapter you should know

- how to sketch the graph of a quadratic function that can be factorised by finding the points of intersection with the x axis and then using the symmetry of the graph to locate the vertex
- how to rewrite $ax^2 + bx + c$ in the form $p(x + q)^2 + r$
- that the vertex of $y = p(x + q)^2 + r$ is at the point $(-q, r)$
- how to use the method of completing the square to sketch the graph of a quadratic function which cannot be factorised
- how to use a sketch graph to solve a quadratic inequality
- that the discriminant, Δ, of the equation $ax^2 + bx + c = 0$ is the number $b^2 - 4ac$ and that if the discriminant is positive then the quadratic equation has two distinct real roots; if the discriminant is zero then the quadratic equation has just one root and if the discriminant is negative then the quadratic equation has no real roots

REVISION EXERCISE

1 Sketch the graphs of

a) $y = 3x^2 + 12x - 8$ **b)** $y = 12 + 4x - x^2$

2 a) i) Calculate the discriminant of the quadratic polynomial $2x^2 + 6x + 7$.

ii) State the number of real roots of the equation $2x^2 + 6x + 7 = 0$, and hence explain why $2x^2 + 6x + 7$ is always positive.

b) The quadratic equation

$$kx^2 + (4k + 1)x + (3k + 1) = 0$$

has a repeated root. Find the value of the constant k.

(OCR Jan 2001 P1)

3 a) Find values p, q and r so that $2x^2 + 6x - 7 = p(x + q)^2 + r$.
b) Write down the equation of the line of symmetry of the curve $y = 2x^2 + 6x - 7$.
c) Write down the co-ordinates of the vertex of the curve $y = 2x^2 + 6x - 7$.

4 Solve the inequalities
a) $4x^2 + 7x \geqslant 30$ **b)** $x^2 \leqslant 6x + 8$

5 The quadratic equation

$$3mx^2 - (m + 3)x + (m - 2) = 0$$

has two real roots.
a) Show that the constant m must satisfy the inequality $11m^2 - 30m - 9 < 0$.
b) Solve this inequality to find the possible values of the constant m.

6 a) i) Express the quadratic polynomial $x^2 - 2\sqrt{2}x + 4$ in the form $(x + a)^2 + b$, stating the exact values of the constants a and b.
ii) Hence write down the equation of the line of symmetry of the curve $y = x^2 - 2\sqrt{2}x + 4$.
b) The quadratic equation $x^2 + (k + 1)x + 16 = 0$ has two distinct real roots. Find the set of possible values of the constant k.

(OCR Jun 2001 P1)

7 Solve the inequality $2x^2 - 10x + 3 > 0$.

8 a) Find constants s, t and u so that $16x^2 - 40x + 7 \equiv (sx + t)^2 + u$.
b) Deduce the co-ordinates of the minimum point of the curve $y = 16x^2 - 40x + 7$.

9 a) Sketch the graph of $y = x^2 - 4\sqrt{5}x - 16$, giving the co-ordinates of the vertex and all points of intersection with the co-ordinate axes.
b) Hence solve the inequality $x^2 > 4\sqrt{5}x + 16$.

10 a) Find constant numbers p, q and r so that $-3x^2 - 12x + 20 \equiv p(x + q)^2 + r$.
b) Hence sketch the graph of $y = -3x^2 - 12x + 20$.
c) Solve the inequality $0 \leqslant -3x^2 - 12x + 20$.

11 i) Express $2x^2 + 4x + 1$ in the form $a((x + p)^2 + q)$, stating the values of the constants a, p and q.
ii) Sketch the graph of $y = 2x^2 + 4x + 1$, stating the co-ordinates of the vertex.

(OCR Jan 2001 P1)

12 a) Sketch the graph of $y = 5x^2 + 6x - 3$.
b) Use your sketch to determine the values of k for which the equation

$$5x^2 + 6x - 3 = k$$

has no real roots.

7 The Gradient of a Curve

The purpose of this chapter is to enable you to

- establish procedures for estimating the gradient of a curve at a point
- discover and prove rules for the gradient functions for curves of the form $y = x^n$

Tangents and Gradients

The tangent to a curve at a point is the line that touches the curve at that point.
The diagram shows the graph of $y = x^2 - 6x$.
The tangent at the point (4, −8) has been drawn onto the diagram.

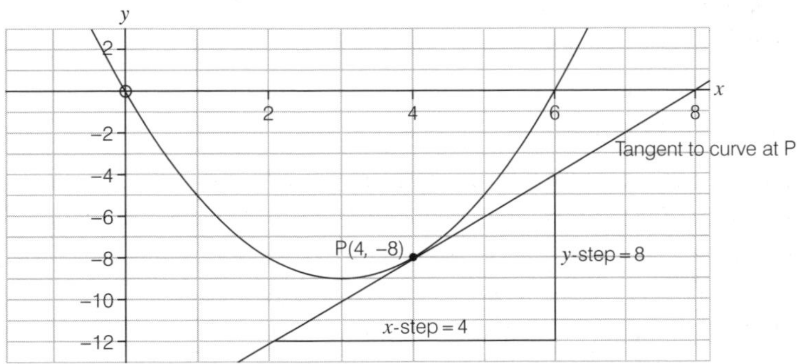

The gradient of this tangent can easily be calculated: gradient $= \dfrac{y \text{ step}}{x \text{ step}} = \dfrac{8}{4} = 2$.

The **gradient of the curve** at a point is defined as the **gradient of the tangent** to the curve at that point.
The example above shows that the gradient of the curve $y = x^2 - 6x$ at the point (4, −8) is 2.

EXERCISE 1

1 The diagram shows an accurate graph of $y = x^2$ for values of x between −4 and 4.
The tangent lines at the points where $x = 1$, 2 and 3 have been drawn in.
Use the diagram to calculate the gradient of $y = x^2$ at the points where $x = 1$, 2 and 3.

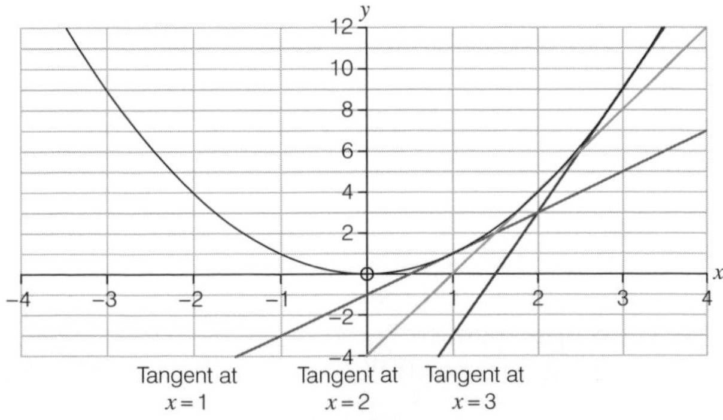

Use the symmetry of the curve to write down the gradient of the curve at the points where $x = -1$, -2 and -3.

Copy and complete the table:

Point	Gradient
(−3, 9)	
(−2, 4)	
(−1, 1)	
(0, 0)	
(1, 1)	
(2, 4)	
(3, 9)	

Write down a rule for the gradient of $y = x^2$ at the point (a, a^2).

The Small Chords Method for Estimating the Gradient of a Curve

Finding the gradient of a graph by drawing the graph is a time-consuming process and it can be extremely inaccurate unless the graph is drawn precisely and the tangent has been drawn very carefully. We need to find methods of finding the gradient of a graph that do not rely on an accurate diagram.

The diagram shows two points P and Q that are close to each other on a curve. The tangent to the curve at P has also been drawn:

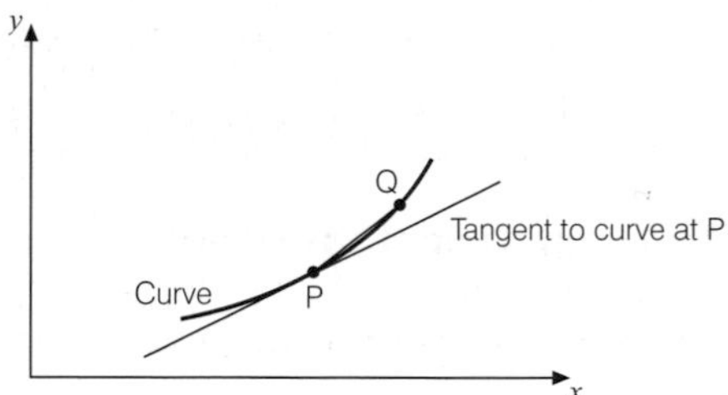

Looking at the diagram we can see that

- if the point Q is reasonably close to the point P then the gradient of the tangent at P may be approximated by the gradient of the chord PQ;
- the closer that Q is to P, the better the approximation will be.

These two observations form the basis of the small chords method for estimating gradients of curves.

Consider the gradient of $y = x^2$ at the point P(3, 9).

To find an estimate of the gradient of $y = x^2$ at the point P(3, 9) you take another point Q just a small way further along the curve and say that the gradient of the curve at P is **approximately** the same as the gradient of the line (or chord) PQ.

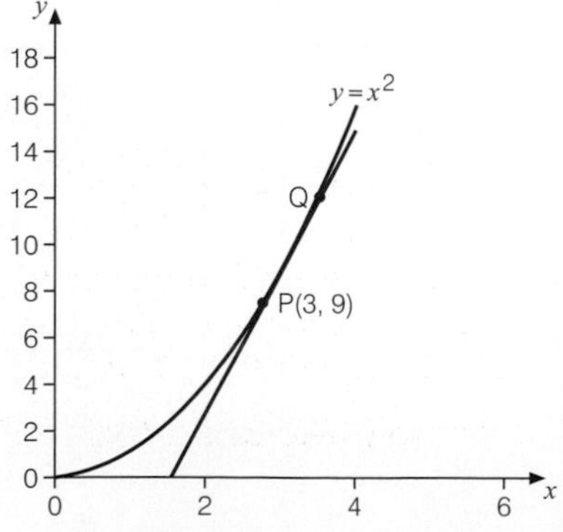

If you take Q to be the point where $x = 3.1$ then the y value of Q is $3.1^2 = 9.61$.

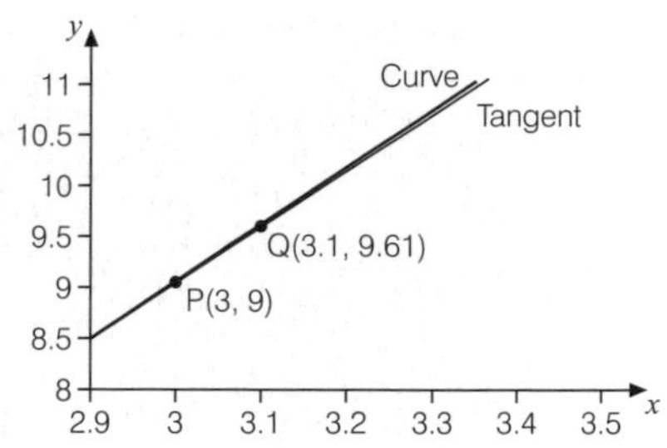

The points P and Q are (3, 9) and (3.1, 9.61), respectively, so

$$\text{gradient of PQ} = \frac{0.61}{0.1} = 6.1$$

so an estimate of the gradient of the curve is 6.1.

This estimate for the gradient of the curve $y = x^2$ at P can be improved by pushing the point Q closer and closer to the point P.

The results can be summarised in a table:

P	Q	Gradient PQ
(3, 9)	(3.1, 9.61)	6.1
(3, 9)	(3.01, 9.0601)	6.01
(3, 9)	(3.001, 9.006001)	6.001
(3, 9)	(3.0001, 9.00060001)	6.0001

$$\text{Grad PQ} = \frac{y \text{ step}}{x \text{ step}} = \frac{0.0601}{0.01} = 6.01.$$

As Q gets closer and closer to P, you can:

- expect the gradient of PQ to get closer and closer to the gradient of the tangent at P;
- see that the gradient of PQ appears to be getting closer and closer to 6.

You can make the **conjecture** that the gradient of $y = x^2$ at the point P(3, 9) is 6.

A conjecture is a result that seems to be true on the basis of evidence that has been gathered but **has not yet been formally proved**.

The gradient of PQ **appears** to be getting closer and closer to 6 but it is not yet certain that this is the case.

To prove this conjecture is correct, you need to work through the small chords method algebraically rather than arithmetically.

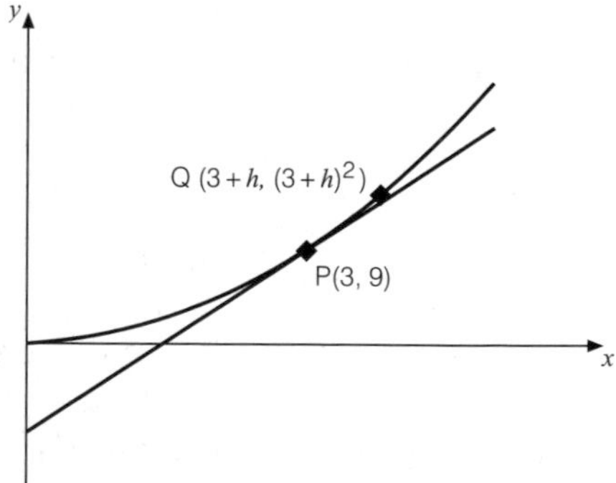

Keeping P as the point (3, 9), let Q be the point on the curve whose x co-ordinate is $3 + h$.

The y co-ordinate of Q will be $(3 + h)^2$ so Q is $(3 + h, (3 + h)^2)$.

$$\text{Gradient PQ} = \frac{y \text{ step}}{x \text{ step}} = \frac{(3+h)^2 - 9}{3 + h - 3} = \frac{9 + 6h + h^2 - 9}{h} = \frac{6h + h^2}{h} = \frac{h(6+h)}{h} = 6 + h.$$

As Q approaches P, the value of h gets closer and closer to zero, so the gradient of PQ does indeed get closer and closer to 6.
This can be written in shorthand: as $Q \rightarrow P$, $h \rightarrow 0$ so Grad $PQ \rightarrow 6$.

It has now been **proved** that the gradient of $y = x^2$ at the point P(3, 9) is 6.

EXAMPLE 1

Use the small chords method with an x step of 0.001 to estimate the gradient of the curve $y = 2x^3 - 5x + 1$ at the point P(2, 7).

SOLUTION

You must first find the co-ordinates of the second point, Q, on the curve.

x co-ordinate of Q = x co-ordinate of P + 0.001 = 2.001.

Since Q lies on the curve $y = 2x^3 - 5x + 1$;

y co-ordinate of Q = $2 \times 2.001^3 - 5 \times 2.001 + 1 = 7.019012002$.

So Q is the point (2.001, 7.019012002).

You can estimate the gradient of the curve at P by saying

gradient of curve at P(2, 7) ≈ gradient of PQ =

$$\frac{\text{y step}}{x \text{ step}} = \frac{0.019012002}{0.001} = 19.012002 \approx 19.$$

EXERCISE 2

1 The small chord method is to be used to investigate the gradient of $y = x^2$ at the point P(6, 36).

a) Complete the following table and use it to make a conjecture about the gradient of $y = x^2$ at the point P(6, 36):

P	Q	Gradient PQ
(6, 36)	(6.1, 37.21)	12.1
(6, 36)	(6.01,)	
(6, 36)	(6.001,)	
(6, 36)	(6.0001,)	

b) Let P be the point (6, 36) and Q be the point $(6 + h, (6 + h)^2)$.
Prove that the gradient of PQ is $12 + h$.
Hence prove that the conjecture made in part (a) is correct.

2 Complete the table below to obtain estimates, using the small chords method, of the gradient of $y = x^2$ at various points.

P	Q	Gradient of PQ	Estimate of curve gradient at P
(1, 1)	(1.001,)		
(2, 4)	(2.001,)		
(3, 9)	(3.001, 9.00601)	6.001	6
(4, 16)	(4.001,)		
(5, 25)	(5.001,)		
(6, 36)	(6.001, 36.012001)	12.001	12

Suggest a rule for the gradient of the curve $y = x^2$ at the point (a, a^2).

3 The points P and Q lie on the curve $y = x^2$ and have coordinates (a, a^2) and $(a + h, (a + h)^2)$, respectively.

a) Show that the gradient of the chord PQ is equal to $2a + h$.

b) What does this tell you about the gradient of $y = x^2$ at the point (a, a^2)?

4 Use the method of small chords with an x step of 0.001 to estimate the gradient of the following curves at the given points:

a) $y = x^3 + 5x$ at $(2, 18)$

b) $y = 5x^2 + 6x - 2$ at $(1, 9)$.

c) $y = 3^x$ at $(2, 9)$

d) $y = \dfrac{24}{x^2}$ at $(2, 6)$

Using a Spreadsheet to Investigate the Gradient of $y = x^n$

So far, you have seen that the small chords method enables you to estimate the gradient of a curve at a point.
You have also seen that the gradient of $y = x^2$ at the point (a, a^2) is $2a$.
This can be written as a rule:

$$y = x^2 \quad \Rightarrow \quad \text{gradient} = 2x.$$

The spreadsheet below has been prepared so that the gradient of $y = x^n$ can be investigated using the small chords method but without having to perform large numbers of calculations.

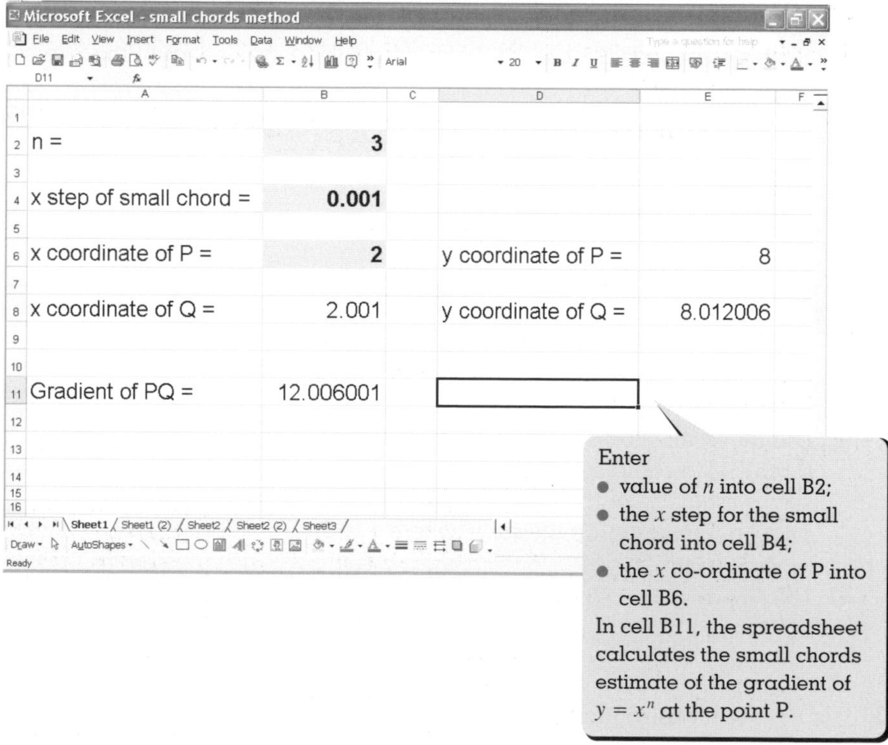

This spreadsheet shows that if the small chords method with an x step of 0.001 is used to estimate the gradient of $y = x^3$ at the point P(2, 8) then an estimate of 12 is obtained.

The screen dump below shows the formulae used in the spreadsheet.

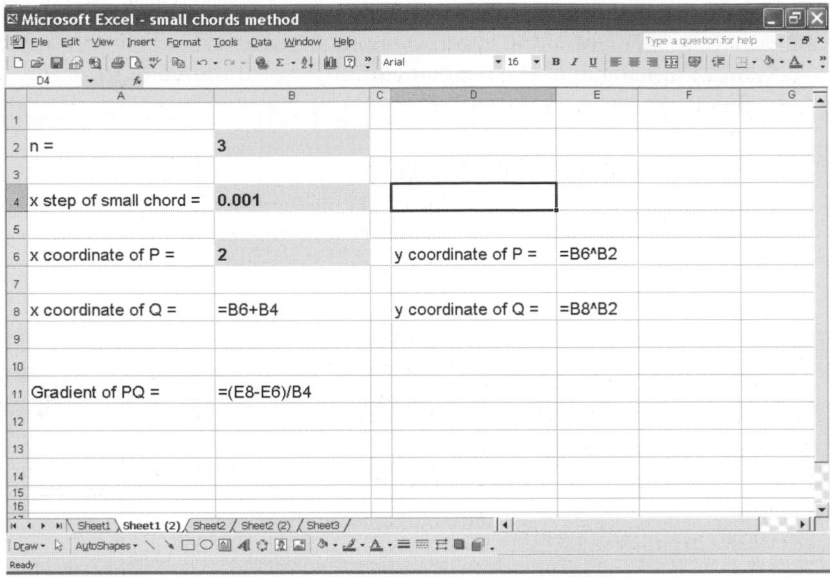

EXERCISE 3

Make sure you work through this exercise before progressing to the next section.

The Excel worksheet shown above could be used for questions 1–6 of this exercise.

1 Consider the curve $y = x^3$.

a) Use the small chords method, with x steps of 0.001, to complete the table so that it shows estimates of the gradient of the curve $y = x^3$ at the points on the curve where x is 0, 1, 2, 3, 4 and 5.

x	Gradient estimate
0	
1	
2	12
3	
4	
5	

b) Use the table to complete the statement

"$y = x^3 \Longrightarrow$ gradient = ..."

2 Consider the curve $y = x^4$.

a) Use the small chords method, with x steps of 0.00001, to complete the table so that it shows estimates of the gradient of the curve $y = x^4$ at the points on the curve where x is 0, 1, 2, 3, 4 and 5.

x	Gradient estimate
0	
1	
2	
3	
4	
5	

b) Use the table to complete the statement

"$y = x^4 \Longrightarrow$ gradient = ..."

3 Use the small chords method, with x steps of 0.00001, to investigate the gradient of $y = x^5$.
Try to find a rule of the form "$y = x^5 \Longrightarrow$ gradient = ..."

4 Consider the curve $y = \sqrt{x}$.

a) Use the small chords method, with x steps of 0.001, to complete the table so that it shows estimates of the gradient of the curve $y = \sqrt{x}$ at the points on the curve where x is 1, 4, 9, 16, 25 and 25.

b) Use the table to complete the statement

"$y = \sqrt{x} \Rightarrow$ gradient = ..."

Hint: try to express the gradient estimates as fractions.

x	Gradient estimate
1	
4	
9	
16	
25	
100	

5 Consider the curve $y = \frac{1}{x}$.

Remember $\frac{1}{x} = x^{-1}$

a) Use the small chords method, with x steps of 0.001, to complete the table so that it shows estimates of the gradient of the curve $y = \frac{1}{x}$ at the points on the curve where x is 1, 2, 3, 4, 5 and 6.

b) Use the table to complete the statement

"$y = \dfrac{1}{x} \Rightarrow$ gradient = ..."

Hint: try to express the gradient estimates as fractions.

x	Gradient estimate
1	
2	
3	
4	
5	
6	

6 Use the small chords method, with x steps of 0.001, to investigate the gradient of $y = \dfrac{1}{x^2}$.

Try to find a rule of the form "$y = \dfrac{1}{x^2} \Rightarrow$ gradient = ..."

7 Summarise your results for questions 1–6 in a table:

Curve	Gradient rule
$y = x^2$	$2x$
$y = x^3$	
$y = x^4$	
$y = x^5$	
$y = x^{0.5} = \sqrt{x}$	
$y = \dfrac{1}{x} = x^{-1}$	
$y = \dfrac{1}{x^2} = x^{-2}$	

Try to find an overall rule for the gradient of $y = x^n$.

Practical Differentiation 1

In the previous sections and in the questions of exercise 3 the following conjectures were obtained:

- the gradient of $y = x^2$ at the point (x, x^2) is $2x$
- the gradient of $y = x^3$ at the point (x, x^3) is $3x^2$
- the gradient of $y = x^4$ at the point (x, x^4) is $4x^3$
- the gradient of $y = \sqrt{x} = x^{0.5}$ at the point $(x, x^{0.5})$ is $0.5x^{-0.5}$

In general:

- the gradient of $y = x^n$ at the point (x, x^n) is nx^{n-1}

This rule agrees with two results you already know about the gradient of straight lines:

You know that the gradient of the line $y = x$ is 1.

The rule predicts that the gradient of $y = x = x^1$ at the point (x, x) should be $1x^{1-1}$, which simplifies to $1x^0$ or 1.

You also know that the gradient of the line $y = 1$ is 0 since it is a horizontal line.

The rule predicts that the gradient of $y = 1 = x^0$ at the point $(x, 1)$ should be $0x^{0-1}$, which simplifies to 0.

Notation:

You write $\frac{\mathrm{d}y}{\mathrm{d}x}$ for the gradient of an x–y curve.

Thus, for example, you will write

$$y = x^3 \quad \Rightarrow \quad \frac{\mathrm{d}y}{\mathrm{d}x} = 3x^2$$

You would read this result out aloud as "If y equals x cubed then dee y dee x equals three x squared".

as shorthand for the statement

"If a curve has equation $y = x^3$ then its gradient at the point (x, x^3) is $3x^2$".

You can say that $\frac{\mathrm{d}y}{\mathrm{d}x}$ is the gradient function of the curve

or that $\frac{\mathrm{d}y}{\mathrm{d}x}$ is the **derivative** of y with respect to x.

The process of finding the derivative of a curve is called **differentiation**.

This basic result may be stated as

$$y = x^n \quad \Rightarrow \quad \frac{\mathrm{d}y}{\mathrm{d}x} = nx^{n-1}$$

Alternative notation:

The gradient of the function $f(x)$ is often written as $f'(x)$.
Thus, for example, $f(x) = x^3 \quad \Rightarrow \quad f'(x) = 3x^2$.

In this notation our basic result states that

$$f(x) = x^n \quad \Rightarrow \quad f'(x) = nx^{n-1}$$

Using the Rule for Gradients in Calculations

The rule for the gradient of x^n gives an extremely quick way of calculating the gradient of curves.

EXAMPLE 2

Calculate the gradient of $y = x^6$ at the point (–2, 64).

SOLUTION

$$y = x^6$$

$$\Rightarrow \quad \frac{dy}{dx} = 6x^5.$$

You want the gradient at the point (–2, 6) where x is –2.

When $x = -2$, gradient $= 6 \times (-2)^5 = 6 \times (-32) = -192$.

EXAMPLE 3

Calculate the gradient of $z = t^2\sqrt{t}$ at the point where $t = 4$.

SOLUTION

$$z = t^2\sqrt{t} = t^2 t^{0.5} = t^{2.5}$$

Note that you need to write z in index form before you try to find the gradient.

$$\Rightarrow \quad \frac{dz}{dt} = 2.5t^{1.5}$$

The gradient of a t, z graph is written $\frac{dz}{dt}$ and the general rule can be written as $z = t^n \Rightarrow \frac{dz}{dt} = nt^{n-1}$.

When $t = 4$, gradient $= 2.5 \times 4^{1.5} = 2.5 \times 8 = 20$.

EXAMPLE 4

If $f(x) = \frac{1}{x^4}$ find the value of $f'(2)$.

SOLUTION

$$f(x) = \frac{1}{x^4} = x^{-4}$$

Again, you must write the function in index form before you try to find its gradient.

$$\Rightarrow \quad f'(x) = -4x^{-5} = -\frac{4}{x^5}$$

$$\Rightarrow \quad f'(2) = -\frac{4}{2^5} = -\frac{4}{32} = -\frac{1}{8}.$$

EXAMPLE 5

Find the derivative of y with respect to x if $y = \dfrac{x^5}{\sqrt{x}}$.

SOLUTION

The question is simply asking you to find $\dfrac{dy}{dx}$:

$$y = \frac{x^5}{\sqrt{x}} = \frac{x^5}{x^{0.5}} = x^{4.5}$$

Again, you must write the function in index form before you try to find its gradient.

$$\Rightarrow \quad \frac{dy}{dx} = 4.5x^{3.5}.$$

EXERCISE 4

1. Find the gradient of $y = x^8$ at the point (1, 1).
2. Find $\dfrac{dy}{dx}$ if $y = x^4\sqrt{x}$.
3. Find $\dfrac{dy}{dt}$ if $y = \dfrac{1}{t^3}$.
4. If $f(x) = x^5$ find the value of $f'(-2)$.
5. Find the gradient of the graph of $z = \sqrt{t}$ at the point where $t = 16$.
6. If $g(t) = \sqrt[3]{t}$ find the value of $g'(8)$.
7. If $y = (p^3)^2$ find $\dfrac{dy}{dp}$.
8. **a)** Write $\dfrac{1}{x\sqrt{x}}$ in the form x^p where p is a rational number.

   b) Find the gradient of the graph $y = \dfrac{1}{x\sqrt{x}}$ at the point $(4, \frac{1}{8})$.
9. **a)** Write $\dfrac{\sqrt{x}}{x^3}$ in the form x^p where p is a rational number.

   b) Find the value of $f'(4)$ if $f(x) = \dfrac{\sqrt{x}}{x^3}$.
10. **a)** Write $(\sqrt[3]{x})^5$ in the form x^p where p is a rational number.

    b) Find $\dfrac{dy}{dx}$ if $y = (\sqrt[3]{x})^5$.
11. **a)** Write $\dfrac{x^2}{(\sqrt{x})^3}$ in the form x^p where p is a rational number.

    b) Find the value of $g'(9)$ if $g(x) = \dfrac{x^2}{(\sqrt{x})^3}$.

12 **a)** Write $\dfrac{t^3}{(\sqrt[3]{t})^4}$ in the form t^p where p is a rational number.

b) Find the gradient of the graph $z = \dfrac{t^3}{(\sqrt[3]{t})^4}$ at the point where $t = 8$.

Using a Spreadsheet to Investigate the Gradient of $f(x) + g(x)$

The spreadsheet used earlier in the chapter can easily be adapted to obtain gradient estimates for curves of the form $y = ax^n + bx^m$.

Microsoft Excel - small chords method

	A	B	C	D	E
1	a =	2		b =	1
2	n =	3		m =	2
3					
4	x step of small chord =	0.001			
5					
6	x coordinate of P =	5		y coordinate of P =	275
7					
8	x coordinate of Q =	5.001		y coordinate of Q =	275.16
9					
10					
11	Gradient of PQ =	160.031			

This spreadsheet shows that if the small chords method with an x step of 0.001 is used to estimate the gradient of $y = 2x^3 + x^2$ at the point P(5, 275) then an estimate of 160 is obtained.

The screen dump below shows the formulae used in this revised spreadsheet.

Microsoft Excel - small chords method

	A	B	C	D	E
1	a =	2		b =	1
2	n =	3		m =	2
3					
4	x step of small chord =	0.001			
5					
6	x coordinate of P =	5		y coordinate of P =	=B1*B6^B2+E1*B6^E2
7					
8	x coordinate of Q =	=B6+B4		y coordinate of Q =	=B1*B8^B2+E1*B8^E2
9					
10					
11	Gradient of PQ =	=(E8-E6)/B4			

EXERCISE 5

Make sure you work through this exercise before progressing to the next section.

The Excel worksheet shown above could be used for questions 1–5 of this exercise.

1 Consider the curve $y = 2x^3 + x^2$.
Use the small chords method, with x steps of 0.001, to complete the table so that it shows estimates of the gradient of the curve $y = 2x^3 + x^2$ at the points on the curve where x is 0, 1, 2, 3, 4 and 5.

What is $\frac{dy}{dx}$ for the curve $y = 2x^3 + x^2$?

x	Gradient estimate
0	
1	
2	
3	
4	
5	160

2 Consider the curve $y = 5x^2 + 7x$.
Use the small chords method, with x steps of 0.001, to complete the table so that it shows estimates of the gradient of the curve $y = 5x^2 + 7x$ at the points on the curve where x is 0, 1, 2, 3, 4 and 5.

What is $\frac{dy}{dx}$ for the curve $y = 5x^2 + 7x$?

x	Gradient estimate
0	
1	
2	
3	
4	
5	

3 Consider the curve $y = x^3 + x^2$.
Use the small chords method, with x steps of 0.001, to complete the table so that it shows estimates of the gradient of the curve $y = x^3 + x^2$ at the points on the curve where x is 0, 1, 2, 3, 4 and 5.

What is $\frac{dy}{dx}$ for the curve $y = x^3 + x^2$?

x	Gradient estimate
0	
1	
2	
3	
4	
5	

4 Consider the curve $y = 2x^2 - 5x$.
Use the small chords method, with x steps of 0.001, to complete the table so that it shows estimates of the gradient of the curve $y = x^3 + x^2$ at the points on the curve where x is 0, 1, 2, 3, 4 and 5.

What is $\frac{dy}{dx}$ for the curve $y = 2x^2 - 5x$?

x	Gradient estimate
0	
1	
2	
3	
4	
5	

5 Consider the curve $y = \frac{36}{x} + 5x$.
Use the small chords method, with x steps of 0.001, to complete the table so that it shows estimates of the gradient of the curve $y = x^3 + x^2$ at the points on the curve where x is 1, 2, 3, 4 and 5.

What is $\frac{dy}{dx}$ for the curve $y = \frac{36}{x} + 5x$?

x	Gradient estimate
1	
2	
3	
4	
5	

6 Summarise your results for questions 1–5 in a table:

Curve	$\frac{dy}{dx}$
$y = 2x^3 + x^2$	
$y = 5x^2 + 7x$	
$y = x^3 + x^2$	
$y = 2x^2 - 5x$	
$y = \frac{36}{x} + 5x$	

Try to find an overall rule for the gradient of $y = ax^n + bx^m$.

Practical Differentiation 2

You have seen in exercise 5 that if

$$y = x^3 + x^2 \quad \text{then} \quad \frac{dy}{dx} = 3x^2 + 2x = \text{gradient of } x^3 + \text{gradient of } x^2$$

and that, if

$$y = 2x^2 - 5x \quad \text{then} \quad \frac{dy}{dx} = 4x - 5 = 2 \times (\text{gradient of } x^2) - 5 \times (\text{gradient of } x).$$

In general, if you have two functions f and g and constant numbers a and b then

$$y = a\,f(x) + b\,g(x) \quad \Rightarrow \quad \frac{dy}{dx} = a\,f'(x) + b\,g'(x)$$

EXAMPLE 6

Find the derivatives of

a) $3x^8 + 6x^2$ **b)** $\frac{3}{x^4}$ **c)** $\frac{1}{2x^2}$ **d)** $(2x + 1)(x^2 - 1)$ **e)** $\frac{x^2 + 1}{2x^3}$

SOLUTION

a) $y = 3x^8 + 6x^2$

$$\Rightarrow \frac{dy}{dx} = 3 \times (\text{gradient of } x^8) + 6 \times (\text{gradient of } x^2) = 3 \times 8x^7 + 6 \times 2x$$

$$\Rightarrow \frac{dy}{dx} = 24x^7 + 12x.$$

b) $y = \frac{3}{x^4} = 3 \times \frac{1}{x^4} = 3x^{-4}$

You must write y in index form before you try to differentiate.

$$\Rightarrow \frac{dy}{dx} = 3 \times (-4x^{-5}) = -12x^{-5} = -12 \times \frac{1}{x^5} = -\frac{12}{x^5}.$$

EXAMPLE 6 (continued)

c) $y = \frac{1}{2x^2} = \frac{1}{2} \times \frac{1}{x^2} = \frac{1}{2}x^{-2}$

Again, it is necessary to write y in index form before you try to differentiate.

$\Rightarrow \frac{dy}{dx} = \frac{1}{2} \times (-2x^{-3}) = -x^{-3} = -\frac{1}{x^3}.$

d) $y = (2x+1)(x^2-1) = 2x^3 + x^2 - 2x - 1$

$\Rightarrow \frac{dy}{dx} = 6x^2 + 2x - 2.$

e) $y = \frac{x^2+1}{2x^3}$

This first stage needs some careful algebra. You must split the expression up into a sum of terms written in index form before you attempt to differentiate.

$= \frac{x^2}{2x^3} + \frac{1}{2x^3}$

$= \frac{1}{2} \times \frac{1}{x} + \frac{1}{2} \times \frac{1}{x^3}$

$= \frac{1}{2}x^{-1} + \frac{1}{2}x^{-3}$

$\Rightarrow \frac{dy}{dx} = -\frac{1}{2}x^{-2} - \frac{3}{2}x^{-4} = -\frac{1}{2x^2} - \frac{3}{2x^4}.$

EXAMPLE 7

Find the gradient of the curve $y = x^2 + \frac{6}{x}$ at the point (1, 7).

SOLUTION

$$y = x^2 + \frac{6}{x} = x^2 + 6x^{-1}$$

$$\Rightarrow \frac{dy}{dx} = 2x - 6x^{-2} = 2x - \frac{6}{x^2}.$$

When $x = 1$,

$$\text{gradient} = 2 \times 1 - \frac{6}{1^2} = 2 - 6 = -4.$$

EXAMPLE 8

Find the points on the curve $y = x^3 - 5x^2 + 4x + 2$ at which the tangent is parallel to the line $y = -4x$.

SOL

The line $y = -4x$ has gradient -4. If the tangent is parallel to this line then the tangent must also have gradient -4. Remembering that the gradient of the tangent at a point is the same as the gradient of the curve at that point, this means that you are looking for the points on the curve $y = x^3 - 5x^2 + 4x + 2$ where the gradient is -4.

EXAMPLE 8 (continued)

Gradient of curve $= \frac{dy}{dx} = 3x^2 - 10x + 4$.

So, you want $3x^2 - 10x + 4 = -4$

$\Rightarrow 3x^2 - 10x + 8 = 0$

$\Rightarrow (3x - 4)(x - 2) = 0$

$\Rightarrow x = 2$ or $\frac{4}{3}$.

The curve has equation $y = x^3 - 5x^2 + 4x + 2$ so

when $x = 2$, $y = 2^3 - 5 \times 2^2 + 4 \times 2 + 2 = 8 - 20 + 8 + 2 = -2$

and when $x = \frac{4}{3}$, $y = (\frac{4}{3})^3 - 5 \times (\frac{4}{3})^2 + 4 \times (\frac{4}{3}) + 2 = \frac{64}{27} - \frac{80}{9} + \frac{16}{3} + 2 = \frac{22}{27}$

so the required points are $(2, -2)$ and $(\frac{4}{3}, \frac{22}{27})$.

EXERCISE 6

1 Find the gradient of each of the following functions:

a) $y = x^3 + 3x + 1$ **b)** $y = 3x^2 + 5x - 7$ **c)** $y = 4x^3 - 5x^2$

d) $y = (3x^2 - 1)(2x + 1)$ **e)** $y = \frac{12}{x^2}$ **f)** $y = 8\sqrt{x}$

g) $y = 5x^2 + \frac{7}{x} - 3$ **h)** $y = \frac{3}{x} - \frac{5}{x^2}$ **i)** $y = 3\sqrt{x} - \frac{4}{\sqrt{x}}$

2 Find $\frac{dy}{dx}$ if

a) $y = 6x^2 - 8x + 3$ **b)** $y = (x - 2)(x^2 + 3x - 3)$

c) $y = 3x^{1.2} - 2x^{0.6}$ **d)** $y = \frac{4x^2 + 1}{x^3}$

3 Find $f'(x)$ if **a)** $f(x) = 3x^2 + 2x - \frac{5}{x}$ **b)** $f(x) = (2x - 1)(x^3 + 5)$

4 Find $\frac{dp}{dt}$ if **a)** $p = 3t^2 - \frac{6}{t}$ **b)** $p = \frac{t^3 - 5t}{t^2}$

5 Find the gradient of the following curves at the given point:

a) $y = x^3 - 5x^2$ at $(5, 0)$ **b)** $y = \frac{16}{x^2}$ at $(4, 1)$

c) $y = 6x - x^2$ at the point where $x = 2$ **d)** $y = x^3$ at the point where $y = 64$

6 Find the coordinates of the points on the curve $y = 3x + \frac{8}{x}$ at which the gradient of the curve is -1.

7 Find the two points on the curve of the function

$$f(x) = 2x^3 - 9x^2 + 12x + 4$$

at which the gradient is zero.

8 At which points on the curve of $y = x^3 - 5x^2$ is the tangent parallel to the line $y = -7x - 21$?

9 At which points on the curve of $y = \frac{4}{x}$ is the tangent perpendicular to the line $y = 25x$?

10 Find $\frac{dy}{dx}$ if

a) $y = \frac{5x - 7}{3x^2}$ **b)** $y = \frac{(x^2 - 1)(x + 3)}{x^3}$ **c)** $y = \frac{3x^2 + 12}{6x}$

EXTENSION

Formalising the Process of Finding the Gradient of a Curve

*This section is **not** required for the C1 examination.*

You have seen how the small chords method can be used arithmetically to obtain estimates of the gradients of curves and that these estimates can lead to conjectures about rules for the gradients of the form $y = x^n$ and $y = ax^m + bx^n$. In this section you will see how the small chords method leads to a formal definition of the gradient of a curve at a point and how the formal definition can then be used to prove gradient results.

Consider the points P(a, f(a)) and Q($a + h$, f($a + h$)) on the curve $y = \mathrm{f}(x)$:

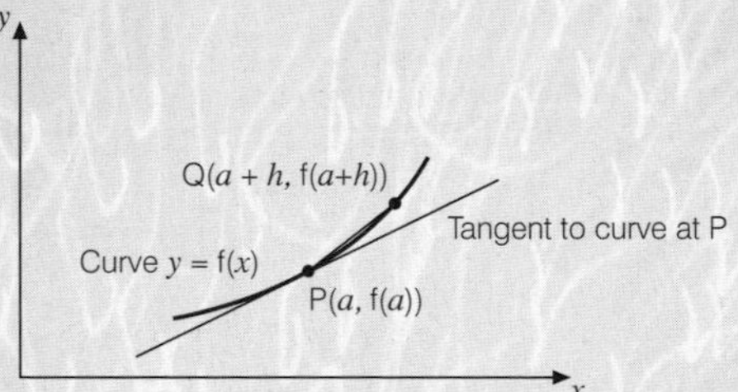

If the value of h is reasonably small then you know that the gradient of the chord PQ gives a reasonable estimate of the gradient of the curve at P. This can be written as

$$\text{gradient of curve at P} \approx \text{gradient PQ}$$
$$\approx \left(\frac{\mathrm{f}(a+h) - \mathrm{f}(a)}{h}\right).$$

Moreover, you know that the closer that Q is to P, the better the estimate will be. This can be written as

$$\text{gradient of curve at P} = \underset{Q \to P}{\text{limit}}(\text{gradient PQ}) = \lim_{h \to 0}\left(\frac{\mathrm{f}(a+h) - \mathrm{f}(a)}{h}\right)$$

and this second gives the formal definition of the gradient, $\mathrm{f}'(a)$, of the curve $y = \mathrm{f}(x)$ at the point where $x = a$:

$$\mathrm{f}'(a) = \lim_{h \to 0}\left(\frac{\mathrm{f}(a+h) - \mathrm{f}(a)}{h}\right).$$

This formal definition should be used to establish proofs of the gradient results discovered earlier in the chapter.

EXAMPLE 9

Prove that the gradient of $y = x^4$ at the point $P(a, a^4)$ is $4a^3$.

SOLUTION

The curve under consideration is $y = f(x) = x^4$.

$$\text{Gradient of curve at P} = \lim_{h \to 0} \left(\frac{f(a+h) - f(a)}{h} \right)$$

$$= \lim_{h \to 0} \left(\frac{(a+h)^4 - a^4}{h} \right)$$

$$\begin{aligned}(a+h)^4 &= (a+h)^2(a+h)^2 \\ &= (a^2 + 2ah + h^2)(a^2 + 2ah + h^2) \\ &= a^4 + 2a^3h + a^2h^2 + 2a^3h + 4a^2h^2 + 2ah^3 + a^2h^2 + 2ah^2 + h^4 \\ &= a^4 + 4a^3h + 6a^2h^2 + 4ah^3 + h^4\end{aligned}$$

$$= \lim_{h \to 0} \left(\frac{a^4 + 4a^3h + 6a^2h^2 + 4ah^3 + h^4 - a^4}{h} \right)$$

$$= \lim_{h \to 0} \left(\frac{4a^3h + 6a^2h^2 + 4ah^3 + h^4}{h} \right)$$

$$= \lim_{h \to 0} \left(\frac{h(4a^3 + 6a^2h + 4ah^2 + h^3)}{h} \right)$$

$$= \lim_{h \to 0} (4a^3 + 6a^2h + 4ah^2 + h^3)$$

$$= 4a^3.$$

As $h \to 0$
$6a^2h \to 0$
$4ah^2 \to 0$
$h^3 \to 0$

EXERCISE 7

1 **a)** Prove that $(a+h)^3 = a^3 + 3a^2h + 3ah^2 + h^3$.

Consider the points $P(a, a^3)$ and $Q(a+h, (a+h)^3)$ on the curve $y = x^3$.

b) Prove that the chord PQ has gradient $3a^2 + 3ah + h^2$.

c) Hence prove that the gradient of the curve $y = x^3$ at the point $P(a, a^3)$ is $3a^2$.

2 Consider the points $P\left(a, \frac{1}{a}\right)$ and $Q\left(a+h, \frac{1}{a+h}\right)$ on the curve $y = \frac{1}{x}$.

a) Prove that the chord PQ has gradient $\frac{-1}{a(a+h)}$.

b) Deduce that the gradient of the curve $y = 1/x$ at the point $P\left(a, \frac{1}{a}\right)$ is $\frac{-1}{a^2}$.

Having studied this chapter you should know how

- to estimate the gradient of a curve by drawing a tangent line and finding the gradient of the tangent
- to estimate the gradient at P arithmetically by using the gradient of a small chord, PQ, where Q is a point on the curve close to P
- to improve this estimate by letting the point Q get closer and closer to the point P
- to use the $\frac{dy}{dx}$ and $f'(x)$ notations for gradient
- to use the rules

$$y = x^n \quad \Rightarrow \quad \frac{dy}{dx} = nx^{n-1}$$

$$y = a\,f(x) + b\,g(x) \quad \Rightarrow \quad \frac{dy}{dx} = a\,f'(x) + b\,g'(x)$$

to find the gradient of curves

REVISION EXERCISE

1 Find $\frac{dy}{dx}$ if

a) $y = 7x^4 - 8x^2 + 9$
b) $y = (5x^2 - 1)^2$
c) $y = (x^3 + 2x - 1)(4x - 3)$
d) $y = 5\sqrt{x} + 7x^2$
e) $y = \frac{16}{\sqrt{x}}$
f) $y = {}^5\sqrt{x}$
g) $y = \frac{x^2 + 5}{x}$
h) $y = \frac{x^4 + 9}{x^2}$
i) $y = \frac{5x^3 + 9x}{3x^2}$
j) $y = \frac{5x - 7}{2x^2}$

2 Find the gradients of the following curves at the given points:

a) $y = x^3 - 5x^2 + 7$ at (2, −5)
b) $y = 8\sqrt{x}$ at (4, 16)
c) $y = (2x^3 + 1)^2$ at (1, 9)
d) $y = \frac{16}{x^2}$ at (2, 4)
e) $y = \frac{5x + 4}{9x^2}$ at (1, 1)
f) $y = \frac{(x - 3)^2}{x^2}$ at (5, 0.16)

3 Find the points on the curve $y = x^3 + 3x^2 - 10x + 4$ where the gradient is 14.

4 Find the point on the curve $y = 8\sqrt{x}$ where the tangent is parallel to the line $y = 2x$.

5 Find the point on the curve $y = x^2 + \frac{16}{x} + 3$ where the gradient is 0.

6 Find the point on the curve $y = \frac{32}{x^2}$ where the tangent is perpendicular to the line $y = 8x$.

7 **a)** Find $\frac{dz}{dt}$ if $z = t^3(3t - 2)$ **b)** Find $g'(u)$ if $g(u) = \frac{4}{u} + 5u + 2$

8 Find the co-ordinates of the points on the curve $y = \frac{1}{2}x^3$ where the gradient is 6.

9 The curve $y = 2x^3 + kx^2 + 5x - 4$ has gradient -11 when $x = 2$. Find the value of the constant k.

10 Find the gradient of the graph $y = 2x^2(x - 4)$ at the point $(-2, -48)$.

11 Find the gradient of the graph of $z = 2t^2 + \frac{32}{\sqrt{t}}$ at the point where $t = 4$.

12 The curve $y = x^2 + \frac{a}{x^2} + b$ passes through the point P(2, 3) and has gradient 1 at P. Find the values of the constants a and b.

8 Further Quadratic Equations

The purpose of this chapter is to enable you to

- recognise equations that can be rewritten as quadratic equations
- solve simultaneous equations where one equation is linear and one is quadratic
- use the discriminant of a quadratic equation to determine the number of points of intersection of a line with the graph of a quadratic function

Equations that are Reducible to Quadratic Equations

You have seen in chapter 3 how to solve quadratic equations using factorisation, completing the square and the quadratic equation formulae. Many equations that appear at first glance to be much more complicated than quadratic equations can, in fact, be rewritten as quadratic equations.

EXAMPLE 1

Solve the equation $u^4 - 11u^2 + 28 = 0$.

Remembering that

$$u^4 = (u^2)^2$$

you see that the equation is really a quadratic equation with u^2 as the unknown.

SOLUTION

Put $x = u^2$ $\quad u^4 - 11u^2 + 28 = 0 \Rightarrow x^2 - 11x + 28 = 0$

$$\begin{aligned}&\Rightarrow (x-4)(x-7) = 0\\&\Rightarrow x = 4 \quad \text{or} \quad x = 7\\&\Rightarrow u^2 = 4 \quad \text{or} \quad u^2 = 7\\&\Rightarrow u = \pm 2 \quad \text{or} \quad u = \pm\sqrt{7}.\end{aligned}$$

EXAMPLE 2

Solve the equation $\dfrac{12}{t^2} + \dfrac{8}{t} = 7$.

SOLUTION

Multiplying through by t^2 to clear the fractions gives

$$\begin{aligned}&12 + 8t = 7t^2\\ \Rightarrow\ &0 = 7t^2 - 8t - 12\\ \Rightarrow\ &0 = (7t+6)(t-2)\\ \Rightarrow\ &t = 2 \quad \text{or} \quad -\tfrac{6}{7}.\end{aligned}$$

EXAMPLE 3

Solve the equation $y^{\frac{2}{3}} - 2y^{\frac{1}{3}} - 15 = 0$.

SOLUTION

$$y^{\frac{2}{3}} - 2y^{\frac{1}{3}} - 15 = 0$$

$\Rightarrow \quad (y^{\frac{1}{3}})^2 - 2y^{\frac{1}{3}} - 15 = 0$ so putting $Y = y^{\frac{1}{3}}$

Remember that $y^{\frac{2}{3}} = (y^{\frac{1}{3}})^2$.

$\Rightarrow \quad Y^2 - 2Y - 15 = 0$

$\Rightarrow \quad (Y-5)(Y+3) = 0$

$\Rightarrow \quad Y = 5$ or $Y = -3$

$\Rightarrow \quad y^{\frac{1}{3}} = 5$ or $y^{\frac{1}{3}} = -3$

$\Rightarrow \quad y = 5^3 = 125$ or $y = (-3)^3 = -27$.

EXAMPLE 4

Solve the equation $x + 7\sqrt{x} - 30 = 0$.

SOLUTION

Putting $t = \sqrt{x}$ means that $t^2 = x$ so you can rewrite the equation as

$$t^2 + 7t - 30 = 0$$

$\Rightarrow \quad (t+10)(t-3) = 0$

$\Rightarrow \quad t = -10$ or $t = 3$.

Now, $t = \sqrt{x}$ and $\sqrt{x}$ is **always a non-negative number** so the $t = -10$ option should be rejected.

$\Rightarrow \quad t = 3$

$\Rightarrow \quad x = t^2 = 9$.

If $t = -10$ had not been rejected you would have obtained a second root: $x = 100$.
Substituting $x = 100$ back into the original equation makes it clear that $x = 100$ is **not** a solution of

$$x + 7\sqrt{x} - 30 = 0.$$

It is always worthwhile checking that your final answers do indeed work in the original equation.

EXAMPLE 5

Solve the equation $\dfrac{3}{x-2} + \dfrac{8}{x+3} = 2$.

SOLUTION

Multiply through by $(x-2)(x+3)$

$\Rightarrow \quad (x-2)(x+3)\dfrac{3}{x-2} + (x-2)(x+3)\dfrac{8}{x+3} = 2(x-2)(x+3)$

$\Rightarrow \quad 3(x+3) + 8(x-2) = 2(x-2)(x+3)$

$\Rightarrow \quad 3x + 9 + 8x - 16 = 2(x^2 + x - 6)$

$\Rightarrow \quad 11x - 7 = 2x^2 + 2x - 12$

$\Rightarrow \quad 0 = 2x^2 - 9x - 5$

$\Rightarrow \quad 0 = (2x+1)(x-5)$

$\Rightarrow \quad 2x + 1 = 0$ or $x - 5 = 0$

$\Rightarrow \quad x = -\frac{1}{2}$ or 5.

EXAMPLE 6

Use the substitution $u = 3^x$ to solve the equation $3^{2x+1} - 28 \times 3^x + 9 = 0$.

SOLUTION

From chapter 1 we know that

$$\begin{aligned} 3^{2x+1} &= 3^{2x} \times 3^1 \\ &= (3^x)^2 \times 3 \\ &= 3u^2 \end{aligned}$$

Since $a^{n+m} = a^n \times a^m$.

Since $a^{np} = (a^p)^n$.

so the equation $3^{2x+1} - 28 \times 3^x + 9 = 0$

can be rewritten as $3u^2 - 28u + 9 = 0$

$$\begin{aligned} &\Rightarrow (3u-1)(u-9) = 0 \\ &\Rightarrow 3u - 1 = 0 \text{ or } u - 9 = 0 \\ &\Rightarrow u = \tfrac{1}{3} \text{ or } u = 9 \\ &\Rightarrow 3^x = \tfrac{1}{3} \text{ or } 3^x = 9 \\ &\Rightarrow x = -1 \text{ or } x = 2. \end{aligned}$$

Since $\frac{1}{3} = 3^{-1}$ and $9 = 3^2$.

EXERCISE 1

Solve the following equations:

1. $u^6 - 7u^3 - 8 = 0$
2. $4t^4 - 37t^2 + 9 = 0$
3. $x - 5\sqrt{x} + 6 = 0$
4. $\dfrac{13}{t^2} - \dfrac{36}{t^4} = 1$
5. $\dfrac{3}{q} + \dfrac{4}{q+1} = 2$
6. $\dfrac{6}{2x-1} + \dfrac{x+5}{3x+1} = 3$
7. $x^{\frac{2}{3}} - x^{\frac{1}{3}} - 6 = 0$
8. $2x^{\frac{1}{2}} - 3x^{\frac{1}{4}} + 1 = 0$
9. $4^{2x} - 5 \times 4^x + 4 = 0$
10. $9^{2x} - 12 \times 9^x + 27 = 0$
11. $5x + 3\sqrt{x} - 26 = 0$
12. $\dfrac{3x}{x+2} = x - 2$
13. $2^{2x+3} - 33 \times 2^x + 4 = 0$
14. $3\sqrt{y} + \dfrac{8}{\sqrt{y}} = 10$

Simultaneous Equations: One Linear, One Quadratic

Consider the problem of finding the points of intersection of the graphs of $y = 5 + 4x - x^2$ and $y + x = 9$.

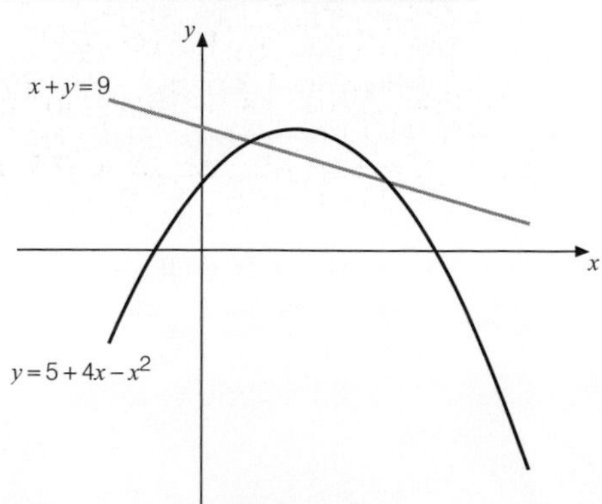

Finding the points of intersection of two graphs is equivalent to solving a pair of simultaneous equations, so you must try to solve

$$\begin{cases} y + x = 9 \\ y = 5 + 4x - x^2 \end{cases}.$$

In this case only one of the equations represents a line; the other represents a curve.

The strategy for solving such simultaneous equations is to use the linear equation to write one of the variables in terms of the other and then substitute this into the other equation.

Rewriting the first equation in a form that gives y in terms of x:

$$y + x = 9 \quad \Rightarrow \quad y = 9 - x.$$

Now substitute $9 - x$ for y in the second equation:

the equation $\quad y = 5 + 4x - x^2$
becomes $\quad 9 - x = 5 + 4x - x^2$
or $\quad x^2 - 5x + 4 = 0.$

Factorising the left hand side gives

$$(x-4)(x-1) = 0$$
$$x - 4 = 0 \quad \text{or} \quad x - 1 = 0$$
$$x = 4 \quad \text{or} \quad x = 1.$$

Finally, find the y values that satisfy the equations.

When $x = 4$: $\quad y = 9 - x = 9 - 4 = 5$
When $x = 1$: $\quad y = 9 - x = 9 - 1 = 8$

So the points of intersection are (4, 5) and (1, 8).

EXAMPLE 7

Solve the simultaneous equations $\begin{cases} 2x - y = 4 \\ x^2 - 5xy + 2y^2 = -13 \end{cases}$.

SOLUTION

Using the first equation to write y in terms of x you get

$$3x - y = 7 \quad \Rightarrow \quad y = 2x - 4.$$

Substituting for y in the second equation:

$$\begin{aligned} x^2 - 5xy + 2y^2 = -13 \quad &\Rightarrow \quad x^2 - 5x(2x-4) + 2(2x-4)^2 = -13 \\ &\Rightarrow \quad x^2 - 10x^2 + 20x + 2(4x^2 - 16x + 16) = -13 \\ &\Rightarrow \quad -x^2 - 12x + 32 = -13 \\ &\Rightarrow \quad 0 = x^2 + 12x - 45 \\ &\Rightarrow \quad (x+15)(x-3) = 0 \\ &\Rightarrow \quad x = 3 \quad \text{or} \quad x = -15. \end{aligned}$$

Using $y = 2x - 4 \quad x = 3 \Rightarrow y = 2$
$\qquad\qquad\qquad\quad x = -15 \Rightarrow y = -34.$

So the solutions of the simultaneous equations are

$$x = 3, y = 2 \quad \text{or} \quad x = -15, y = -34.$$

Always make sure you give your final answer in a format that makes it clear which x value belongs with which y value!

EXERCISE 2

Solve

1 $\left.\begin{aligned} y &= 7x - 10 \\ y &= x^2 \end{aligned}\right\}$ **2** $\left.\begin{aligned} x + y &= 3 \\ 4x - y^2 &= 0 \end{aligned}\right\}$ **3** $\left.\begin{aligned} x + y &= 4 \\ x^2 + y^2 &= 10 \end{aligned}\right\}$

4 $\left.\begin{aligned} y &= 4x + 3 \\ y &= x^2 - 3x - 5 \end{aligned}\right\}$ **5** $\left.\begin{aligned} p + q &= 3 \\ p^2 - q^2 &= 15 \end{aligned}\right\}$ **6** $\left.\begin{aligned} 11x - y &= 19 \\ y &= 2x^2 - 3x + 1 \end{aligned}\right\}$

7 The T shape in the diagram has a perimeter of 40 cm and an area of 36 cm^2. Determine the values of x and y.

8 Find all possible solutions of the simultaneous equations

$p + 2q = 3 \qquad p^2 + 3pq + q^2 = -1$

9 Find all possible solutions of the simultaneous equations

$2x - y = 7 \qquad x^2 - 2xy + 2y^2 = 13$

10 Find all possible solutions of the simultaneous equations

$x - 3y = 8 \qquad 2x^2 - 5xy - 3y^2 = 16$

11 Find the points of intersection of the line $y + 2x = 4$ with the curve $y^2 = 4x$.

12 The diagram shows the curve $2x^2 + y^2 = 54$ and the line $y + 4x = 18$.
Find the co-ordinates of the points of intersection of the line and the curve.

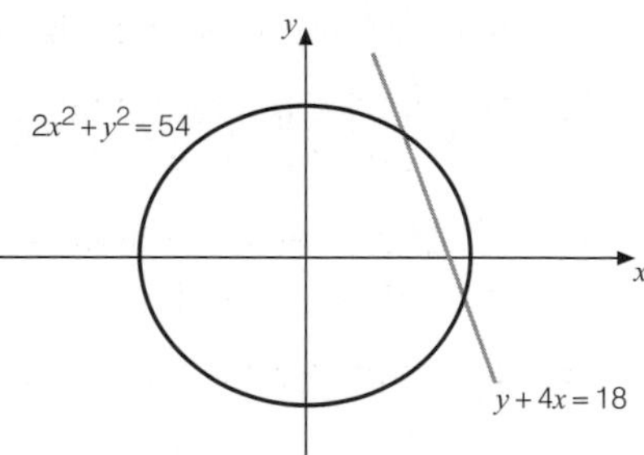

Using the Discriminant

Recall that the **discriminant**, Δ, of the equation $ax^2 + bx + c = 0$ is given by $\Delta = b^2 - 4ac$ and that:

if $\Delta > 0$ then the equation has two different real roots;
if $\Delta = 0$ then the equation has just one real root;
if $\Delta < 0$ then the equation has no real roots.

The diagram shows that the curve of a quadratic polynomial may intersect a straight line, of finite gradient, at 0, 1 or 2 points and that **if there is only one point of intersection then the line is a tangent to the curve.**

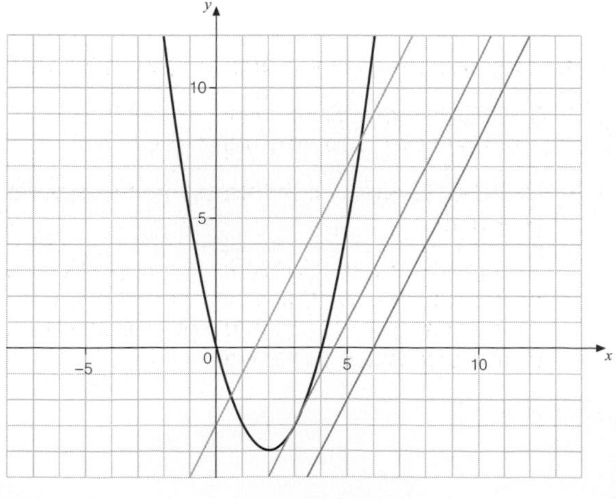

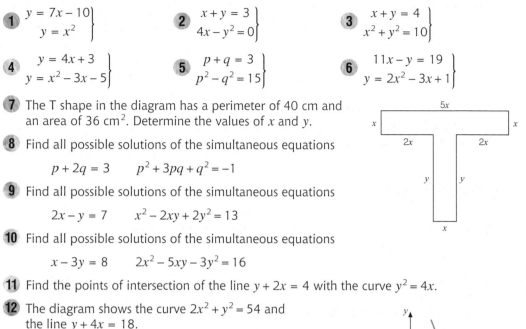

It is possible to use the **discriminant** to tell you how many points of intersection there will be between the curve of a quadratic polynomial and a straight line.

EXAMPLE 8

Determine the number of points of intersection of the curve $y = 2x^2 + 5x - 3$ and the line $y = 9x - 5$.

SOLUTION

$$\left.\begin{aligned} y &= 2x^2 + 5x - 3 \\ y &= 7x - 1 \end{aligned}\right\} \Rightarrow \quad 2x^2 + 5x - 3 = 9x - 5 \quad \Rightarrow \quad 2x^2 - 4x + 2 = 0.$$

For this equation $\Delta = b^2 - 4ac = (-4)^2 - 4 \times 2 \times 2 = 0$ so there is just one solution to the equation and therefore just one point of intersection between the line and curve. **The line must be a tangent to the curve.**

EXERCISE 3

1 **a)** Show that the curve $y = 3x^2 + 5x + 7$ does not meet the x axis.
b) How many points of intersection are there for the curve $y = 3x^2 + 5x + 7$ and the line $y = 2x + 13$?

2 How many points of intersection are there for the curve $y = 3x^2 + 5$ and the line $y = 6x + 2$?
What is the geometrical significance of your answer?

3 **a)** Find the points of intersection of the line $y = 3x + 14$ with the curve $y = x^2 - 3x + 7$.
b) Prove that the line $y = 3x - 2$ is a tangent to the curve $y = x^2 - 3x + 7$.

4 Prove that the line $y = mx - 2$ intersects the curve $y = x^2$ in two distinct points provided $m > \sqrt{8}$ or $m < -\sqrt{8}$.

5 **a)** Prove that the line $y = 5x - 2$ intersects the curve $y = 2x^2 - 7x + 1$ in two distinct points.
b) Find the value of c for which the line $y = 5x + c$ is a tangent to the curve $y = 2x^2 - 7x + 1$.

6 Prove that the line $y + x = -2$ is a tangent to the curve $y^2 = 8x$.

Having studied this chapter you should know how

- to rewrite equations such as $t^4 - 9t^2 + 8 = 0$ and $u^{\frac{2}{5}} - 3u^{\frac{1}{5}} + 2 = 0$ as quadratic equations and hence solve them
- to solve a pair of simultaneous equations where one equation is linear and one is quadratic
- to use the discriminant of a quadratic equation to determine the number of points of intersection of a line with a quadratic function and, in particular, realise that if there is only one point of intersection then the line is a tangent to the curve of the quadratic function

REVISION EXERCISE

1 Solve the equations

a) $y^4 - 2y^2 - 8 = 0$ **b)** $t^6 + 7t^3 - 8 = 0$ **c)** $2p^{\frac{2}{3}} - 7p^{\frac{1}{3}} + 3 = 0$

2 Find the points of intersection of the curve $y = 2x^2 + 5x - 9$ with the line $y = 9x + 7$.

3 **a)** **i)** Calculate the discriminant of $5x^2 - 3x + 7$.

ii) Write down the number of points of intersection of the curve $y = 5x^2 - 3x + 7$ with the x axis.

b) Show that the line $y = 7x + 2$ is a tangent to the curve $y = 5x^2 - 3x + 7$.

4 Solve the equation $3x - 11\sqrt{x} - 20 = 0$.

5 Find the exact values of all the roots of the equation $x^4 - 10x^2 + 24 = 0$.

6 By putting $t = \sqrt{x}$, show that the equation $\sqrt{x} + \dfrac{12}{\sqrt{x}} = 7$

can be written as $t^2 - 7t + 12 = 0$ and hence solve the equation $\sqrt{x} + \dfrac{12}{\sqrt{x}} = 7$.

7 It is given that x and y satisfy the simultaneous equations

$$y - 2x = 2 \qquad x^2 + 3xy - y^2 = 4$$

a) Show that $3x^2 - 2x - 8 = 0$.

b) Hence solve the simultaneous equations.

8 **i)** Solve the simultaneous equations $\left.\begin{array}{r} y = 2x + 2 \\ y = x^2 + 3x - 18 \end{array}\right\}$.

ii) Show that the simultaneous equations $\left.\begin{array}{r} y = 2x - 20 \\ y = x^2 + 3x - 18 \end{array}\right\}$ have no real solutions.

iii) The graph of $y = 2x + k$ meets the graph of $y = x^2 + 3x - 18$ at only one point. Find the value of the constant k.

(OCR Jan 2002 P1)

9 The diagram shows the graphs of $y = 2x^2 - 25$ and $y = 4x - 9$

a) Find the co-ordinates of the points of intersection of the line and the curve.

b) Hence, or otherwise, solve the inequality $2x^2 - 25 < 4x - 9$.

c) The line $y = 4x - k$ does not meet the curve $y = 2x^2 - 25$. Prove that $k > 27$.

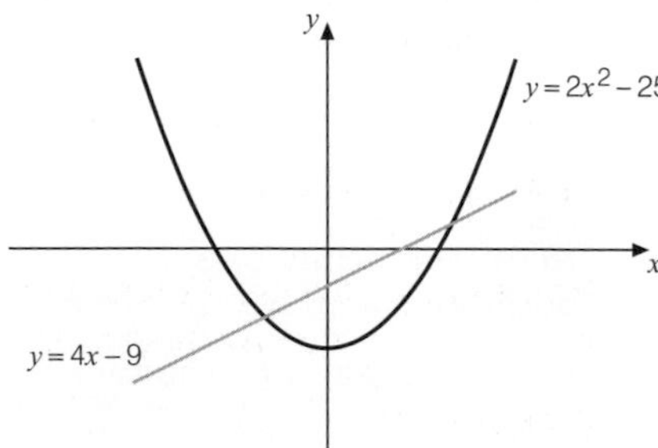

9 Applications of Differentiation

The purpose of this chapter is to enable you to

- recognise the connection between rate of change and gradient
- understand that an increasing function has a positive gradient and that a decreasing function has a negative gradient
- find stationary points of a curve and classify them as a maximum point or a minimum point
- use differentiation to solve optimisation problems
- find the equation of the tangent and normal to a curve at a point
- find and use the second derivative of a function

Rates of Change

You have already seen that $\frac{dy}{dx}$ gives the gradient of a graph of y against x and another way of describing this is as **the rate of change of y with respect to x**.

If the velocity, v, of a particle at time t is given by the formula $v = 12t^2 - 2t^3$ then

$$\frac{dv}{dt} = 24t - 6t^2.$$

This can be described as
the gradient of a velocity–time graph
or as the rate of change of velocity with respect to time
or as the acceleration of the particle.

Similarly, you know that the formula $V = \frac{4}{3}\pi r^3$ gives the volume of a sphere of radius r.

In this case $\frac{dV}{dr} = 4\pi r^2$.

This denotes the gradient of a graph of volume against radius
or the rate of change of volume with respect to radius.

EXAMPLE 1

A box has a square base of side $2x$ cm and height $(x + 2)$ cm. The volume of the box is V cm^3. Find the rate of change of the volume with respect to x when $x = 3$ cm.

SOL

$$\begin{aligned} V &= 2x \times 2x \times (x+2) \\ &= 4x^2(x+2) \\ &= 4x^3 + 8x^2. \end{aligned}$$

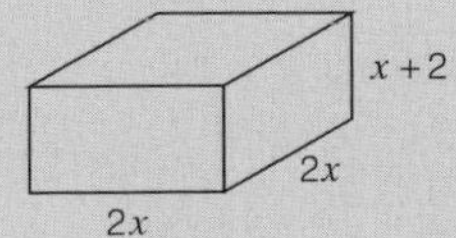

EXAMPLE 1 (continued)

Rate of change of volume with respect to $x = \dfrac{dV}{dx} = 12x^2 + 16x$.

So when $x = 3$,
rate of change of volume with respect to $x = 12 \times 3^2 + 16 \times 3 = 108 + 48 = 156$.

Increasing and Decreasing Functions

The diagram shows the graph of $y = x^3 - 6x^2 + 3$.

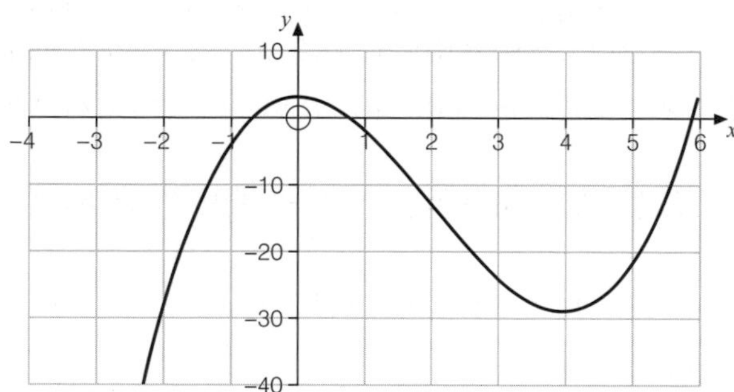

For negative values of x, the function is **increasing** and the gradient of the curve is **positive**; between $x = 0$ and $x = 4$ the function is **decreasing** and the gradient of the curve is **negative**; when $x > 4$ the function is **increasing** and the gradient of the curve is **positive**.

EXAMPLE 2

Find the values of x for which the function $f(x) = x^3 - 3x^2 - 45x + 7$ is increasing.

SOLUTION

If the function f(x) is increasing then the gradient, f′(x), must be positive.

You want $f'(x) = 3x^2 - 6x - 45 > 0$

$\Rightarrow \quad 3(x^2 - 2x - 15) > 0$

$\Rightarrow \quad x^2 - 2x - 15 > 0.$

It is possible to solve this inequality by drawing a sketch of the graph of

$$y = x^2 - 2x - 15 = (x - 5)(x + 3).$$

The graph will

- be ∪ shaped
- cross the x-axis at (5, 0) and (–3, 0).

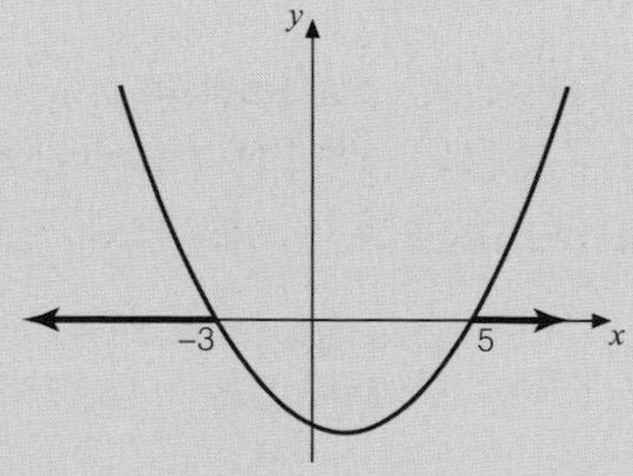

From the graph it can be seen that

$$x^2 - 2x - 15 > 0$$

when $x < -3$ or when $x > 5$.

The function f is increasing when $x < -3$ or when $x > 5$.

EXERCISE 1

1 If $V = 0.2t^3 + 0.6t^2 + 0.5t + 1.2$ find the rate of change of V when $t = 1.2$.

2 If $S = \pi r^2 + \frac{600}{r}$ find the rate of change of S with respect to r when $r = 5$.

3 An oil slick is circular and, t minutes after the initial spillage, the slick has a radius of $(0.5 + 0.2t)$ m.

a) Write down an expression for the circumference, C, of the oil slick at time t.

b) Find $\frac{dC}{dt}$ and explain its significance.

c) Write down an expression for the area, A, of the oil slick at time t.

d) Show that the rate of change of the area when $t = 3$ is approximately 1.38 m^2/min.

4 Determine the values of x for which the function $g(x) = x^2 + 8x - 10$ is a decreasing function.

5 Determine the values of x for which the function $h(x) = x^2 - 18x - 12$ is an increasing function.

6 Determine the values of x for which the function $m(x) = 2x^3 + 4x^2 - 8x + 5$ is an increasing function.

7 Determine the values of x for which the function $n(x) = -4x^3 + 7x^2 - 4x + 5$ is an increasing function.

Stationary Points

It has already been shown that knowledge of the maximum or minimum points of a quadratic function greatly helps when you want to sketch a graph of the function.

For other functions too, knowledge of the maximum and minimum points is often a key step in producing a sketch of the graph of the function.

For example, suppose that you know that the graph of $y = g(x)$ has the following properties:

- the graph is **continuous**, i.e. there are no sudden jumps in the graph
- the graph has one maximum point which is at (1, 7)
- the graph has two minimum points which are at (5, –3) and (–8, –22)
- as $x \to \infty$, $y \to \infty$; this is shorthand for "as x takes larger and larger positive values, the value of y also gets larger and larger without any limit"
- as $x \to -\infty$, $y \to 0$; this is shorthand for "as x takes larger and larger negative values, the value of y gets closer and closer to 0"

then it is possible to show this information on a diagram

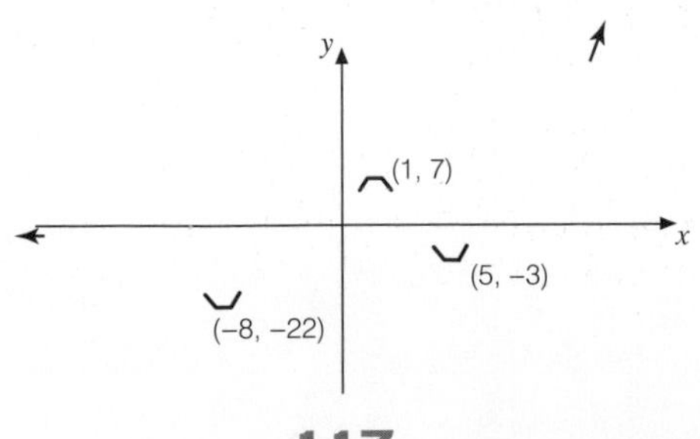

and, since these are the only maximum and minimum points and the graph is continuous there is only really one way of completing the sketch graph:

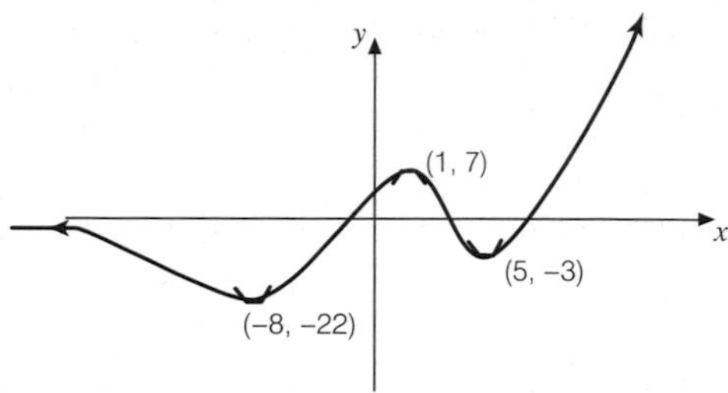

Since knowledge of maximum and minimum points is so important in producing sketch graphs, it is necessary to have a method of finding these points. Fortunately, differentiation provides such a method since the gradient of a curve at a maximum or minimum point must be 0.

A **stationary point** of a curve is a point where the gradient of the curve is 0.
In other words, **a stationary point is a point where**

$$\frac{\mathrm{d}y}{\mathrm{d}x} = 0.$$

A stationary point can take any one of four different forms shown in the diagram.

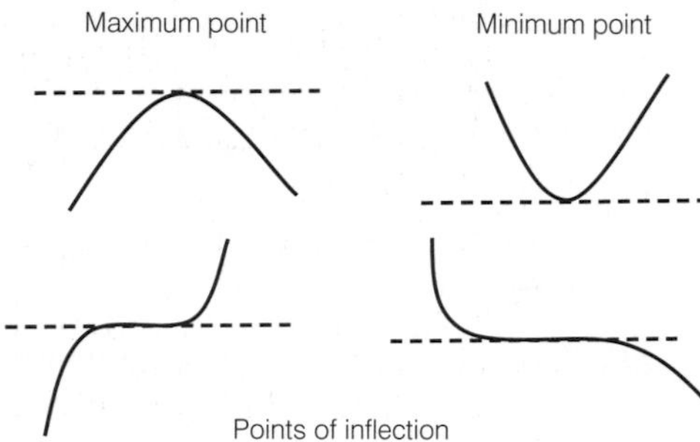

Notice that, in moving from left to right through a stationary point

- **If the gradient is positive, then 0, then negative then the stationary point must be a maximum.**
- **If the gradient is negative, then 0, then positive then the stationary point must be a minimum.**
- **If the gradient is negative, then 0, then negative or if the gradient is positive, then 0, then positive then the stationary point must be a point of inflection.**

EXAMPLE 3

Being asked to "classify the stationary points" is an instruction asking you to decide whether the stationary point is a maximum point, a minimum point or a point of inflection.

Find the stationary points of the curve $y = x^3 + 6x^2 - 36x + 60$ and classify them.
Hence sketch the graph and write down the number of roots of the equations

a) $x^2 + 6x^2 - 36x + 60 = 0$
b) $x^3 + 6x^2 - 36x + 60 = 200.$

SOLUTION

To find the stationary points of a curve we must first find the gradient of the curve.

$$y = x^3 + 6x^2 - 36x + 60$$

$$\Rightarrow \quad \frac{\mathrm{d}y}{\mathrm{d}x} = 3x^2 + 12x - 36 = 3(x^2 + 4x - 12) = 3(x + 6)(x - 2).$$

The factorised form for $\frac{\mathrm{d}y}{\mathrm{d}x}$ is very useful in the working to follow.

EXAMPLE 3 (continued)

The stationary points are where $\frac{dy}{dx} = 0$:

$$\Rightarrow \quad 3(x+6)(x-2) = 0$$
$$\Rightarrow \quad x = -6 \quad \text{or} \quad x = 2.$$

Recalling that the equation of the curve is $y = x^3 + 6x^2 - 36x + 60$,

when $x = -6$, $y = (-6)^3 + 6(-6)^2 - 36(-6) + 60 = -216 + 216 + 216 + 60 = 276$
and
when $x = 2$, $y = 2^3 + 6 \times 2^2 - 36 \times 2 + 60 = 8 + 24 - 72 + 60 = 20$
so the stationary points are (–6, 276) and (2, 20).

The nature of the stationary points can easily be determined by finding whether the gradient is positive or negative either side of the stationary point:

x	$x < -6$	-6	$-6 < x < 2$	2	$2 < x$
y		276		20	
$\frac{dy}{dx} = 3(x+6)(x-2)$	+	0	–	0	+

If $x < -6$ then $(x + 6)$ is negative and $(x - 2)$ is negative so $3(x + 6)(x - 2)$ will be positive.

If $-6 < x < 2$ then $(x + 6)$ is positive and $(x - 2)$ is negative so $3(x + 6)(x - 2)$ will be negative.

If $2 < x$ then $(x + 6)$ is positive and $(x - 2)$ is positive so $3(x + 6)(x - 2)$ will be positive.

As we move from left to right through the stationary point $x = -6$, the gradient goes from positive to 0 to negative.

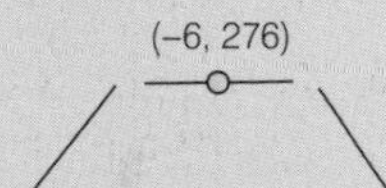

The point (–6, 276) is therefore a maximum point.

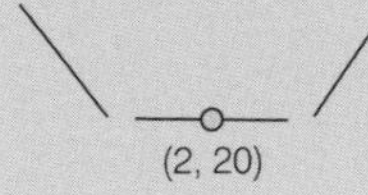

As we move from left to right through the stationary point $x = 2$, the gradient goes from negative to 0 to positive.

The point (2, 20) is therefore a minimum point.

A sketch of the graph can now be produced.

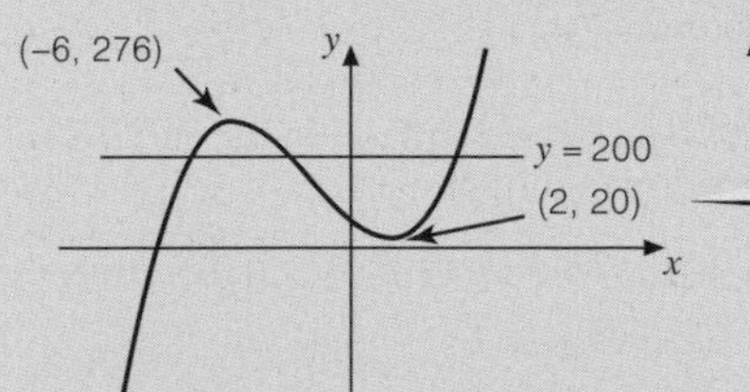

Your sketch should indicate what happens for large positive and large negative values of x.
In this example, for large values of x the x^3 term is the most important so

- if x is large and positive then y will be large and positive;
- if x is large and negative then y will be large and negative.

Looking at the sketch graph you can see that the curve crosses the line $y = 0$ in just one place so the equation $x^3 + 6x^2 - 36x + 60 = 0$ has just one root.

You can also see that the curve crosses the line $y = 200$ in three places, so the equation $x^3 + 6x^2 - 36x + 60 = 200$ has three roots.

EXERCISE 2

In questions 1–5, find the stationary points of the given curves. Classify these stationary points and hence sketch the curve:

1 $y = x^2 - 6x$

2 $y = 3x - x^3$

3 $y = x(x^3 - 12)$

4 $y = x^3 - 3x^2 - 24x + 3$

5 $y = x^2(18 - x^2)$

6 **a)** Solve the equation $x^4 - 50x^2 + 49 = 0$.
b) Find and classify the stationary points of the curve $y = x^4 - 50x^2 + 49$.
c) Sketch the graph of $y = x^4 - 50x^2 + 49$.
d) Write down the number of roots of the equations
i) $x^4 - 50x^2 + 49 = 20$
ii) $x^4 - 50x^2 + 49 = -500$
iii) $x^4 - 50x^2 + 49 = -600$

7 **a)** Find and classify the stationary points of the curve $y = x^3 - 9x^2 + 24x$.
b) Sketch the curve of $y = x^3 - 9x^2 + 24x$.
c) For which values of the constant k does the equation $x^3 - 9x^2 + 24x = k$ have three distinct roots?

8 Consider the curve $y = x^2 + \frac{16}{x}$.
a) Explain why there is no y value when $x = 0$.
b) When x is a small positive number (such as 0.01), y is a large positive number. What happens when x is a small negative number (such as -0.01)?
c) When x is a large positive number (such as 100), y is a large positive number. What happens when x is a large negative number (such as -100)?
d) Find and classify the stationary point of the curve.
e) Sketch the curve.

9 Consider the curve $y = x + \frac{9}{x}$.
a) Explain why there is no y value when $x = 0$.
b) When x is a small positive number (such as 0.01), y is a large positive number. What happens when x is a small negative number (such as -0.01)?
c) When x is a large positive number (such as 100), y is a large positive number. What happens when x is a large negative number (such as -100)?
d) Find and classify the stationary point of the curve.
e) Sketch the curve.

10 Find the stationary point of the curve $y = 1 + \frac{1}{x} - \frac{2}{x^2}$ and classify it.

Sketch the curve.

11 Find the stationary points of the curve $y = 2x^{\frac{7}{5}} - 7x^{\frac{6}{5}}$ and classify them as maximum or minimum points.

Optimisation Problems

The fact that the gradient of a graph is zero at a maximum or minimum point is very useful in finding analytical solutions to a wide range of practical optimisation problems.

EXAMPLE 4

From a rectangular sheet of card measuring 30 cm by 20 cm, four congruent squares are cut from the corners and the remaining sheet is folded up to form an open box. Determine the size of square that should be cut in order to obtain the maximum possible volume.

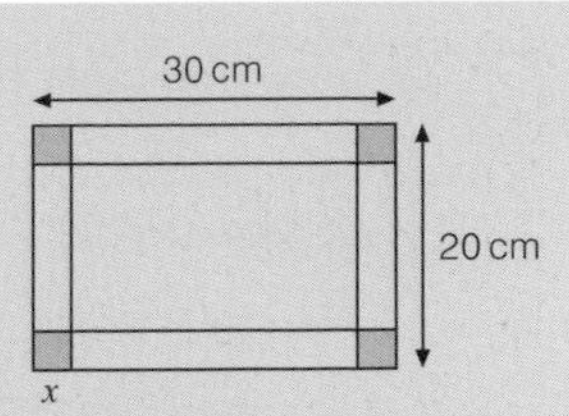

SOLUTION

If a square of side x cm is cut from each corner then the box formed has

length $= 30 - 2x$ width $= 20 - 2x$ height $= x$.

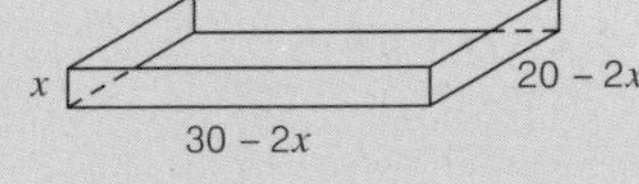

If the volume of the box is V cm^3 then

$$V = (30 - 2x)(20 - 2x)x$$
$$= (600 - 100x + 4x^2)x$$
$$= 4x^3 - 100x^2 + 600x.$$

It is possible to draw a graph to show the relationship between x and V.

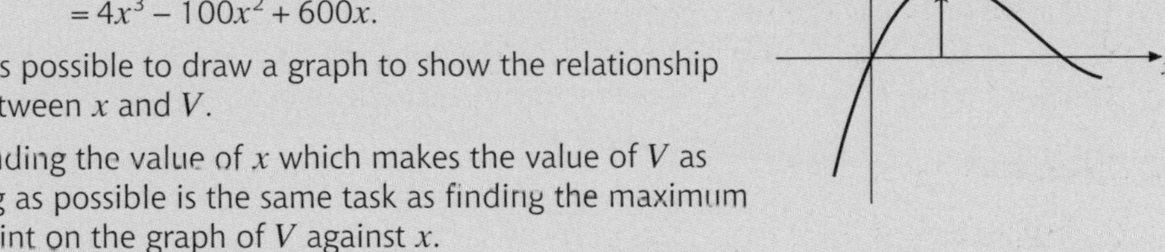

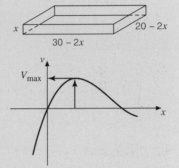

Finding the value of x which makes the value of V as big as possible is the same task as finding the maximum point on the graph of V against x.

You must first find $\frac{dV}{dx}$:

$$V = 4x^3 - 100x^2 + 600x \quad \Rightarrow \quad \frac{dV}{dx} = 12x^2 - 200x + 600.$$

At a maximum point $\frac{dV}{dx} = 0$

$$\Rightarrow \quad 12x^2 - 200x + 600 = 0$$
$$\Rightarrow \quad 3x^2 - 50x + 150 = 0$$
$$\Rightarrow \quad x = \frac{50 \pm \sqrt{50^2 - 4 \times 3 \times 150}}{6} = 3.9237\ldots \quad \text{or} \quad 12.743\ldots$$

Since the base of the box has width $20 - 2x$, the corner size cannot exceed 10 cm so the optimum corner size appears to be 3.9237 ... cm.

> You have not yet proved that the graph has a maximum at $x = 3.9237\ldots$
> All that you have discovered so far is that the graph has a stationary point at $x = 3.9237\ldots$

You should now check that this really does give a maximum by considering the gradient either side of 3.9237 ...

x	$x < 3.9237\ldots$	$3.9237\ldots$	$x > 3.9237\ldots$
$\frac{dV}{dx} = 12x^2 - 200x + 600$	+	0	–

> For example, if $x = 3.9$ $\frac{dV}{dx} = 2.52$.

> For example, if $x = 3.94$ $\frac{dV}{dx} = -1.7168$.

EXAMPLE 4 (continued)

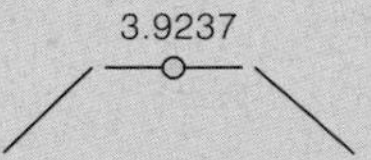

Passing through the stationary point from left to right, the gradient goes from positive to 0 to negative so the stationary point must be a maximum.

You now know that the volume is a maximum when $x = 3.9237 \ldots$

Returning to the formula $V = x(30 - 2x)(20 - 2x)$
it is easy to calculate the maximum possible volume for the box:

$$\begin{aligned} V_{\max} &= 3.9237\ldots \times (30 - 2 \times 3.9237\ldots)(20 - 2 \times 3.9237\ldots) \\ &= 1056.3 \text{ cm}^3 \quad \text{(to one decimal place).} \end{aligned}$$

EXAMPLE 5

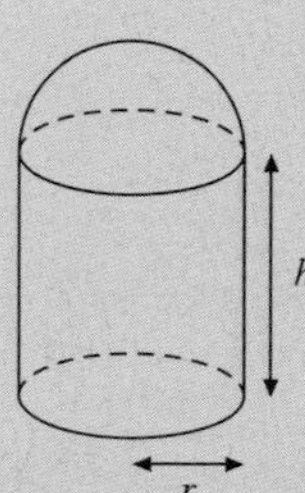

A corn silo is to be constructed in the shape of a cylinder capped by a hemisphere. The floor, sides and roof of the silo are to be constructed from metal sheet. The volume of the silo is to be 360π m^3.
What dimensions should the silo have if the amount of metal used in its construction is to be minimised?

SOLUTION

The dimensions of the silo are completely determined by the radius, r, and height, h, of the cylinder.

The volume of the silo must be 360π m^3.
This can be used to obtain a relationship linking r and h:

$$\begin{aligned} & & 360\pi &= \text{volume of cylinder} + \text{volume of hemisphere} \\ & \Rightarrow & 360\pi &= \pi r^2 h + \tfrac{2}{3}\pi r^3 \\ [\times 3/\pi] & \Rightarrow & 1080 &= 3r^2 h + 2r^3 \\ & \Rightarrow & 3r^2 h &= 1080 - 2r^3 \\ & \Rightarrow & h &= \frac{1080 - 2r^3}{3r^2}. \end{aligned}$$

Recall that:
volume of sphere = $\frac{4}{3}\pi r^3$
surface area of sphere = $4\pi r^2$

The total surface area, A, of the silo gives an indication of the amount of metal that must be used in its construction.

$$\begin{aligned} & A = \text{area of floor} + \text{curved surface area of cylinder} + \text{surface area of hemisphere} \\ \Rightarrow\ & A = \pi r^2 + 2\pi r h + 2\pi r^2 \\ \Rightarrow\ & A = 3\pi r^2 + 2\pi r \frac{1080 - 2r^3}{3r^2} \\ \Rightarrow\ & A = 3\pi r^2 + \frac{2160\pi r}{3r^2} - \frac{4\pi r^4}{3r^2} \\ \Rightarrow\ & A = 3\pi r^2 + \frac{720\pi}{r} - \frac{4\pi r^2}{3} \\ \Rightarrow\ & A = \frac{5}{3}\pi r^2 + \frac{720\pi}{r}. \end{aligned}$$

EXAMPLE 5 (continued)

You want to find the value of r which makes the value of A as small as possible – that is you want to find the minimum point on the graph of A against r.

Now $A = \frac{5}{3}\pi r^2 + \frac{720\pi}{r} \quad \Rightarrow \quad \frac{dA}{dr} = \frac{10}{3}\pi r - \frac{720\pi}{r^2}$.

At a stationary point $\frac{dA}{dr} = 0 \quad \Rightarrow \quad \frac{10}{3}\pi r - \frac{720\pi}{r^2} = 0$

$$\Rightarrow \quad \frac{10}{3}\pi r = \frac{720\pi}{r^2}$$

$$\Rightarrow \quad r^3 = 216$$

$$\Rightarrow \quad r = 6.$$

You should now check that this stationary point is a minimum point:

r	$r < 6$	$r = 6$	$r > 6$
$\frac{dA}{dr} = \frac{10}{3}\pi r - \frac{720\pi}{r^2}$	–	0	+

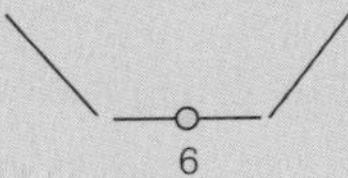

Passing through the stationary point from left to right, the gradient goes from negative to 0 to positive so the stationary point must be a minimum.

The smallest area of metal is obtained when $r = 6$.

Using the formulae $h = \frac{1080 - 2r^3}{3r^2}$ and $A = \frac{5}{3}\pi r^2 + \frac{720\pi}{r}$ with $r = 6$, we obtain

$$h = \frac{1080 - 432}{3 \times 36} = 6 \qquad A = \frac{5}{3}\pi \times 36 + \frac{720\pi}{6} = 180\pi.$$

The silo should be built so that the cylinder has a radius of 6 m and a height of 6 m, which gives a minimum surface for the silo area of 180π m^3.

EXERCISE 3

1 A length of fencing 600 m long is used to make a rectangular enclosure against a long stone wall.

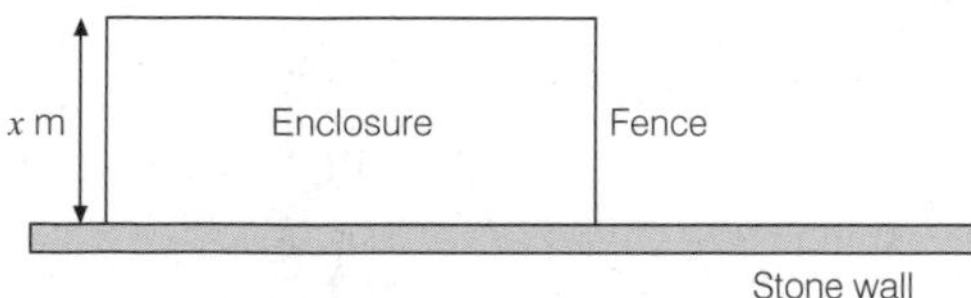

If the enclosure is x metres wide and has an area of A m^2:

i) Prove that $A = 600x - 2x^2$;

ii) Differentiate A with respect to x and hence find the dimensions of the enclosure when A is a maximum. Show clearly that, in this case, S is a maximum and not a minimum.

2 A cylindrical can is to be made to hold 440 cm^3 of drink.
Suppose that the can has radius r cm and height h cm. Write down an equation that must be satisfied by r and h and make h the subject of this equation.

Show that the total surface area, S cm^2, of the can is given by $S = \dfrac{880}{r} + 2\pi r^2$.

Find the values of r and h which minimise the value of S.
Why would this be useful to the manufacturer?
Compare your answer with reality: what assumptions have you made — how realistic are they?

3 The diagram shows a lawn in the shape of a rectangle with a semi-circle at one end.
The lawn must have an area of 200 m^2.

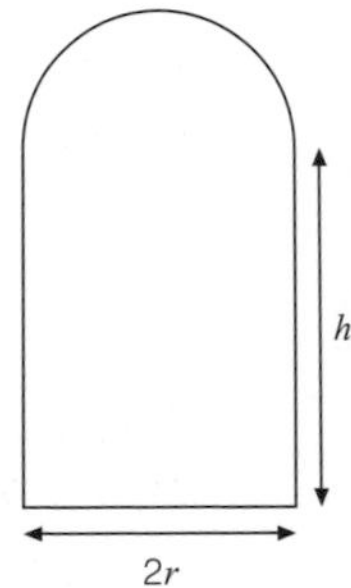

a) Prove that $h = \dfrac{100}{r} - \dfrac{1}{4}\pi r$.

b) If P is the perimeter of the lawn, prove that

$$P = \frac{200}{r} + 2r + \frac{1}{2}\pi r.$$

c) Find $\dfrac{\mathrm{d}P}{\mathrm{d}r}$ and hence find the value of r that will minimise the value of P.

4 A square based box of volume 5000 cm^3 is to be made using a framework of rods.
The horizontal rods cost 5p/cm whilst the vertical rods cost 8p/cm.
Show that if the base has sides of length 25 cm then the total cost of the framework for the box will be £12.56.

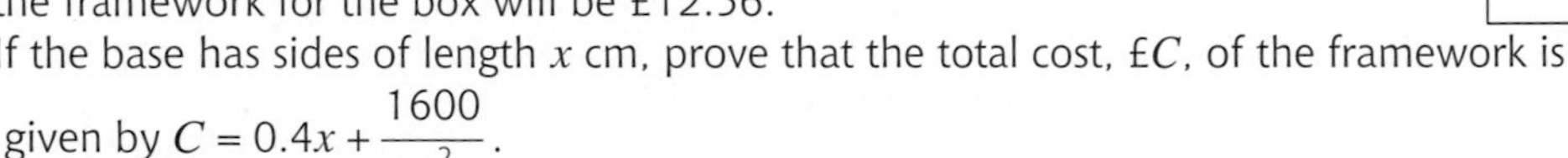
If the base has sides of length x cm, prove that the total cost, £C, of the framework is given by $C = 0.4x + \dfrac{1600}{x^2}$.

Find the value of x that minimises the value of C.

5 A cycle manufacturing company has designed a new bicycle which is about to be put into production. The board has met to discuss the selling price of the new bike.

It is thought that the selling price, £s, will strongly influence the monthly demand, n, for the bike. The Marketing Director produces this graph to show how he thinks the two variables will be linked.

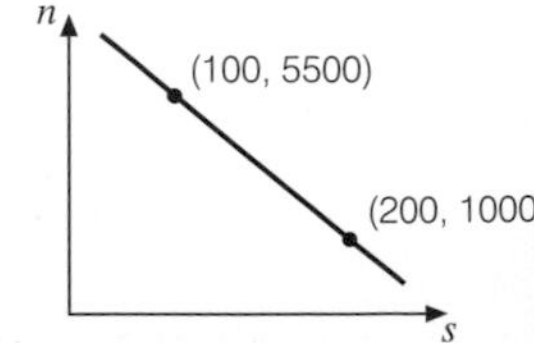

a) Find the equation which links n to s.

The Production Director estimates that the cost of production will be £50 per bike plus capital costs of £60 000 per month to cover the costs of research, machinery, etc.

b) Write down an equation that links the monthly running costs, £C, to the number of bikes, N, that are produced during the month.

c) Show that if the board sets a price of £s as the selling price for the bike and decides to make the number of bikes that will exactly meet the demand then the monthly profit, £P, is given by

$$P = -45s^2 + 12\,250s - 560\,000.$$

d) Determine the selling price that the board should set in order to maximise their profit. How many bikes should they make each month?

6 A ferry crosses from Porthaven to Bridgetown, a distance of approximately 120 km, and after 2 hours at Bridgetown returns to Porthaven.

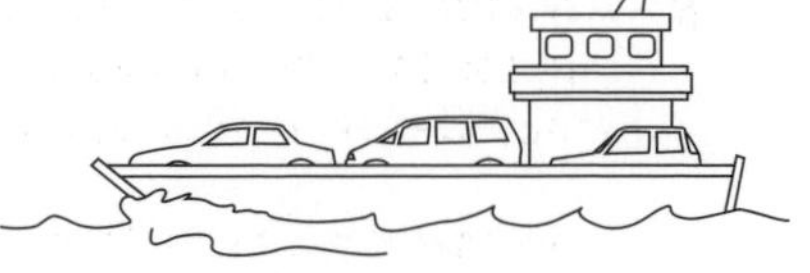

The ferry must pay a fee of £700 to enter Bridgetown harbour and £400 to enter Porthaven.

A total of 64 people work on the ferry and the average earnings of these people are £10.24 per hour. All these people are based in Porthaven; their shift starts one hour before the ferry leaves Porthaven and finishes one hour after the ferry returns to Porthaven.

The fuel for the ferry costs £0.4 per litre. The fuel consumption of the ferry varies with the speed of the ferry as shown in the table:

v km/hr	10	20	30	40
Fuel consumption (litres/km)	16	32	48	64

Wear and tear on the ferry is estimated to cost £4.5 per km travelled.

a) Show that if the ferry travels at 20 km/hr then the total cost of the return crossing Porthaven ⟶ Bridgetown ⟶ Porthaven is £15 737.76

b) At what speed should the ferry travel in order to minimise the total cost of this return journey?

Tangents and Normals

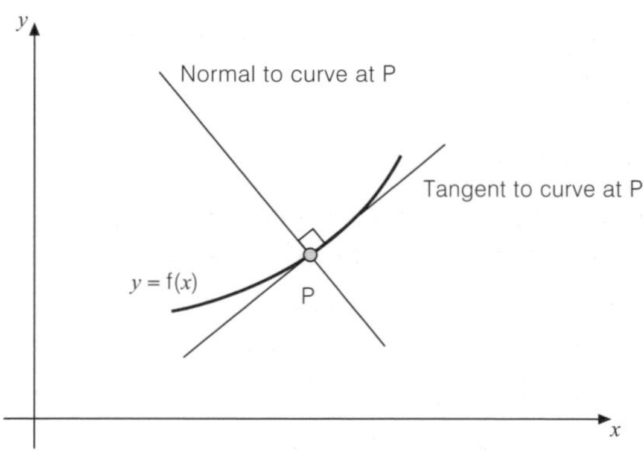

The tangent to a curve at a point P is the line that touches the curve at P. The gradient of the tangent is the same as the gradient of the curve.

The normal to a curve at P is a line that passes through P and is perpendicular to the tangent at P.

Since the tangent and normal are perpendicular you know that

gradient of tangent × gradient of normal = −1.

EXAMPLE 6

Find the equation of the tangent and normal to the curve $y = x^3$ at the point P(2, 8).

SOLUTION

$$y = x^3 \Rightarrow \frac{dy}{dx} = 3x^2 \quad \Rightarrow \quad \text{gradient at P(2, 8)} = 12.$$

The tangent has gradient 12 and passes through (2, 8) so the equation of the tangent is $y - 8 = 12(x - 2)$

Remember that the line with gradient m through the point (x_1, y_1) has equation $y - y_1 = m(x - x_1)$.

$$\Rightarrow \quad y - 8 = 12x - 24$$

$$\Rightarrow \quad y = 12x - 16.$$

The normal has gradient $-\frac{1}{12}$ and passes through (2, 8) so the equation of the normal is $y - 8 = -\frac{1}{12}(x - 2)$.

Remember that

$$\text{gradient of normal} = \frac{-1}{\text{gradient of tangent}}.$$

$$\Rightarrow \quad y - 8 = -\tfrac{1}{12}x + \tfrac{1}{6}$$

$$\Rightarrow \quad y = -\tfrac{1}{12}x + \tfrac{49}{6} \quad \text{or} \quad 12y + x = 98.$$

EXAMPLE 7

Find the equation of the normal to the curve $y = \dfrac{16}{x}$ at the point P(2, 8).

Sketch the curve and the normal.
Determine the co-ordinates of the point Q, distinct from P, where the normal meets the curve again.

SOLUTION

$$y = \frac{16}{x} \quad \Rightarrow \quad \frac{dy}{dx} = -\frac{16}{x^2} \quad \Rightarrow \quad \text{gradient of curve at } x = 2 \text{ is } -\frac{16}{4} = -4$$

$$\Rightarrow \quad \text{gradient of normal} = \frac{-1}{-4} = \frac{1}{4}.$$

The normal has gradient $\frac{1}{4}$ and passes through (2, 8)

so the equation of the normal is $y - 8 = \frac{1}{4}(x - 2)$

$$\Rightarrow \quad y - 8 = \tfrac{1}{4}x - \tfrac{1}{2}$$

$$\Rightarrow \quad y = \tfrac{1}{4}x + \tfrac{15}{2} \quad \text{or} \quad 4y - x = 30.$$

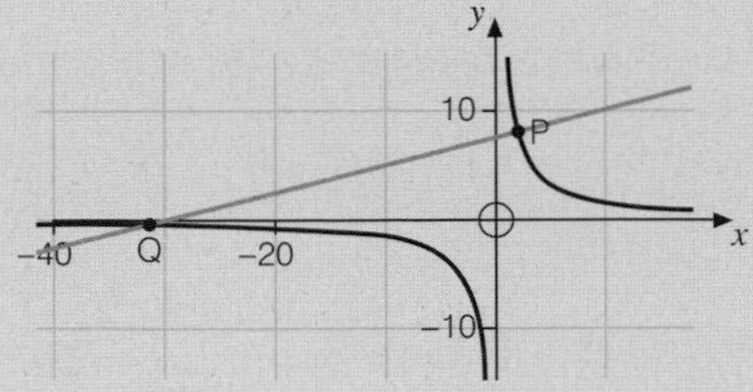

To find the point Q we must solve the simultaneous equations:

$$\begin{cases} y = \dfrac{16}{x} \\ y = \dfrac{1}{4}x + \dfrac{15}{2} \end{cases}$$

EXAMPLE 7 (continued)

$$\Rightarrow \quad \frac{16}{x} = \frac{1}{4}x + \frac{15}{2}$$

$$[\times 4x] \quad \Rightarrow \quad 64 = x^2 + 30x$$

$$\Rightarrow \quad 0 = x^2 + 30x - 64$$

$$\Rightarrow \quad 0 = (x-2)(x+32)$$

$$\Rightarrow \quad x = 2 \quad \text{or} \quad -32$$

At Q, $\quad x = -32 \quad \text{so } y = \frac{16}{-32} = \frac{-1}{2}$

Q is the point $\left(-32, \frac{-1}{2}\right)$.

EXERCISE 4

1 Find the equations of the tangents and normals of the following curves at the given points:
a) $y = x^2$ at (2, 4) **b)** $y = 2x^3 - 4x + 7$ at (1, 5)

2 **a)** Prove that the tangent to the curve $y = \frac{8}{x^2}$ at $(-4, \frac{1}{2})$ has equation $4y - x = 6$.

b) Prove that the normal to the curve $y = (3x - 1)(x - 2)$ at (2, 0) has equation $5y + x - 2 = 0$.

3 Find the equations of the tangent and normal to the curve $y = x^2(x - 3)$ at the point A(3, 0). The tangent meets the y axis at the point T. The normal meets the y axis at the point N. Prove that the triangle ATN has area 41 units.

4 The tangent to the curve $y = \frac{12}{x}$ at the point (2, 6) meets the x axis at P and the y axis at Q. Prove that the triangle OPQ has area 24 units.

5 Find the equations of the normals to the parabola $y = \frac{x^2}{4}$ at the points (−2, 1) and (−4, 4). Show that the point of intersection of these lines also lies on the parabola. Draw a sketch to illustrate your answer.

6 Find the co-ordinates of the points on the curve $y = x^3 - 3x^2 + 2$ at which y has a maximum or a minimum value and distinguish clearly between these points.
Show that the curve passes through the point A(3, 2).
Find the equation of the tangent to the curve at A and the co-ordinates of the point where this line crosses the y axis.

7 Find the equation of the tangent to the curve $y = 0.1x^3 - 0.6x^2$ at the point A(5, −2.5). The point B also lies on the curve and the tangent at B is parallel to the tangent at A. Calculate the co-ordinates of the point B.

8 **a)** Prove that the normal to the curve $y = \frac{16}{x}$ at the point (8, 2) has equation $y = 4x - 30$.

b) The normal to the curve $y = \frac{16}{x}$ at the point (8, 2) is also the tangent to the curve $y = x^2 + k$ at the point $(a, a^2 + k)$. Determine the values of the integers a and k.

9 Prove that the tangent to the curve $y = x^2$ at the point P(p, p^2) has equation $y = 2px - p^2$ and find the equation of the normal.
The tangent meets the y axis at T and the normal meets the y axis at N. M is the midpoint of TN. Find the co-ordinates of M and check that they are independent of the value of p. Draw a sketch to illustrate your answer.

10 Prove that the gradient of $y = x + 6x^{\frac{1}{3}}$ at the point (8, 20) is $\frac{3}{2}$. Find the equation of the normal to the curve $y = x + 6x^{\frac{1}{3}}$ at the point (8, 20), giving your answer in the form $ax + by + c = 0$.

The Second Derivative

If $y = x^3 + 5x^2 - 5$ then $\frac{dy}{dx} = 3x^2 + 10x$

and there is nothing to prevent us from differentiating this again to obtain the second derivative of y which is usually written as:

$$\frac{d^2y}{dx^2} = \frac{d}{dx}\left(\frac{dy}{dx}\right) = \frac{d}{dx}(3x^2 + 10x) = 6x + 10.$$

This result would be read out aloud as "dee two y dee x squared equals six x plus ten".

If you are using function notation then the second derivative of f(x) is usually written as f″(x).

Thus, if

$$f(x) = 5x^2 + 3 - \frac{10}{x}$$

then $f'(x) = 10x + \frac{10}{x^2}$ and $f''(x) = 10 - \frac{20}{x^3}$.

EXERCISE 5

1 Find $\frac{d^2y}{dx^2}$ if

a) $y = 4x^5 + 7x^2 - 15$ **b)** $y = 3x^4 + \frac{5}{x}$ **c)** $y = 5\sqrt{x}$

d) $y = \left(x^3 + \frac{1}{x^2}\right)^2$ **e)** $y = \frac{3x - 2}{x^2}$

2 **a)** Find $\frac{d^2p}{dt^2}$ if $p = 2t^3 + \frac{5}{t^2}$.

b) Find g″(4) if g$(x) = 12\sqrt{x}$.

3 For each of the following curves

i) find the co-ordinates of the stationary points and classify them

ii) find the value of $\frac{d^2y}{dx^2}$ at each of the stationary points:

a) $y = x^2 - 8x$; **b)** $y = 10x - x^2$ **c)** $y = x^3(x - 3)$ **d)** $y = x^4 - 8x^2 + 7$

What can you say about the nature of a stationary point if the value of $\frac{d^2y}{dx^2}$ at the stationary point is positive?

What can you say about the nature of a stationary point if the value of $\frac{d^2y}{dx^2}$ at the stationary point is negative?

Using the Second Derivative to Determine the Nature of a Stationary Point

Suppose that $x = a$ is a stationary point on the curve $y = f(x)$ and that $f''(a) > 0$.

$\Rightarrow$ The gradient of $y = f'(x)$ at $x = a$ is positive.

Remember that $f''(x)$ is the result of differentiating $f'(x)$: in other words, $f''(x)$ is the gradient of $f'(x)$.

$\Rightarrow$ $f'(x)$ is an **increasing** function near $x = a$.

Remember that if the gradient is positive then the function is increasing.

$\Rightarrow$ If x is a little bigger than a then $f'(x)$ is a little bigger than $f'(a)$ which is 0, i.e. $f'(x) > 0$.

$x = a$ is a stationary point on the curve $y = f(x)$ so $f'(a) = 0$.

If x is a little smaller than a then $f'(x)$ is a little smaller than 0, i.e. $f'(x) < 0$.

x	Just smaller than a	a	Just larger than a
Gradient	\	— • —	/

So, passing through the stationary point from left to right, the gradient goes from negative to 0 to positive and the stationary point must therefore be a minimum.

If $\frac{dy}{dx} = 0$ and $\frac{d^2y}{dx^2} > 0$ at $x = a$ then $x = a$ is a minimum point.

In a similar way we can prove that

If $\frac{dy}{dx} = 0$ and $\frac{d^2y}{dx^2} < 0$ at $x = a$ then $x = a$ is a minimum point.

These two results give an alternative method for deciding the nature of most stationary points: **calculate the value of $\frac{d^2y}{dx^2}$ at the stationary point, if the result is positive then the stationary point is a minimum point and if the result is negative then the stationary point is a maximum**.

For many examples, when $\frac{d^2y}{dx^2}$ is easy to calculate, this will provide a quicker means of determining the nature of the stationary point.

However, if the value of $\frac{d^2y}{dx^2}$ at the stationary point is zero, then the method cannot be used. In the Extension section at the end of the chapter you will see that a stationary point where $\frac{d^2y}{dx^2} = 0$ could be a maximum point or a minimum point or a point of inflection.

If $\frac{d^2y}{dx^2} = 0$ at a stationary point then we must use the method of considering the sign of $\frac{dy}{dx}$ either side of the stationary point to determine the nature of the stationary point.

EXAMPLE 8

Find the stationary point on the curve $y = x^2 + \frac{16}{x} - 5$ and classify it as a maximum or minimum point.

SOLUTION

$$y = x^2 + \frac{16}{x} - 5 = x^2 + 16x^{-1} - 5$$

$$\Rightarrow \quad \frac{dy}{dx} = 2x - 16x^{-2} = 2x - \frac{16}{x^2}.$$

At a stationary point $\frac{dy}{dx} = 0$

$$\Rightarrow \quad 2x - \frac{16}{x^2} = 0$$

$$\Rightarrow \quad 2x = \frac{16}{x^2}$$

$$\Rightarrow \quad 2x^3 = 16$$

$$\Rightarrow \quad x^3 = 8$$

$$\Rightarrow \quad x = 2 \quad \text{and} \quad y = x^2 + \frac{16}{x} - 5 = 4 + 8 - 5 = 7$$

so the stationary point is (2, 7).

Now $\quad \frac{dy}{dx} = 2x - 16x^{-2}$

$$\Rightarrow \quad \frac{d^2y}{dx^2} = 2 + 32x^{-3} = 2 + \frac{32}{x^3}.$$

When $x = 2$, $\quad \frac{dy}{dx} = 0 \quad$ and $\quad \frac{d^2y}{dx^2} = 6 > 0$.

You now know that the curve has a **minimum point** at (2, 7).

EXERCISE 6

1 If $y = x^2 + 12x - 28$

a) find the values of x for which $y = 0$

b) find the co-ordinates of the point P where $\dfrac{dy}{dx} = 0$

c) find the value of $\dfrac{d^2y}{dx^2}$ at the point P and hence classify P as a maximum or minimum point

d) sketch the graph of $y = x^2 + 12x - 28$.

2 If $y = x + \dfrac{9}{x} - 10$

a) find the values of x for which $y = 0$

b) find the co-ordinates of the points P and Q where $\dfrac{dy}{dx} = 0$

c) find the value of $\dfrac{d^2y}{dx^2}$ at the points P and Q and hence classify P and Q as maximum or minimum points

d) sketch the graph of $y = x + \dfrac{9}{x} - 10$.

3 Find, and classify, the stationary points of the curve $y = 8x^2 - x^4$.
Hence sketch the curve $y = 8x^2 - x^4$.
For what values of the constant k does the equation $8x^2 - x^4 = k$ have four real roots?

4 Find and classify the stationary points of the curve $y = x^3 + \dfrac{48}{x} - 5$.

5 Prove that if $x = b$ is a stationary point on the curve $y = f(x)$ and $f''(b) < 0$ then the stationary point is a maximum.

EXTENSION

Working with Points of Inflection

A detailed knowledge of points of inflection is not required by the C1 specification but a discussion of stationary points is incomplete without giving points of inflection some attention.

EXAMPLE 9

Find the stationary points of the curve $y = x^3(x - 4)$ and classify them. Hence sketch the curve.

SOLUTION

To find the stationary points of a curve you must first find the gradient of the curve:

$$y = x^3(x - 4) = x^4 - 4x^3$$

$$\Rightarrow \quad \frac{dy}{dx} = 4x^3 - 12x^2 = 4x^2(x - 3).$$

EXAMPLE 9 (continued)

The stationary points are where $\frac{dy}{dx} = 0$:

$$\begin{aligned} & 4x^2(x-3) = 0 \\ \Rightarrow\quad & x^2 = 0 \quad \text{or} \quad x - 3 = 0 \\ \Rightarrow\quad & x = 0 \quad \text{or} \quad x = 3. \end{aligned}$$

Now $y = x^3(x - 4)$ so
when $x = 0$, $y = 0^3(0 - 4) = 0$
and when $x = 3$, $y = 3^3(3 - 4) = -27$.

So the stationary points are (0, 0) and (3, −27).

The nature of the stationary points can easily be determined by finding whether the gradient is positive or negative either side of the stationary point:

x	$x < 0$	0	$0 < x < 3$	3	$3 < x$
y		0		−27	
$\frac{dy}{dx} = 4x^2(x-3)$	−	0	−	0	+

If $x < 0$ then x^2 is positive and $(x - 3)$ is negative so $4x^2(x - 3)$ will be negative.

If $0 < x < 3$ then x^2 is positive and $(x - 3)$ is negative so $4x^2(x - 3)$ will be negative.

If $x > 3$ then x^2 is positive and $(x - 3)$ is positive so $4x^2(x - 3)$ will be positive.

As we move from left to right through the stationary point $x = 0$, the gradient goes from negative to 0 to negative.

(0, 0)

The point (0, 0) is therefore a point of inflection.

As we move from left to right through the stationary point $x = 3$, the gradient goes from negative to 0 to positive.

(3, −27)

The point (3, −27) is therefore a minimum point.

Using the knowledge gained from the stationary points, together with the fact that it is clear that $y = 0$ when $x = 4$, we can now sketch the curve:

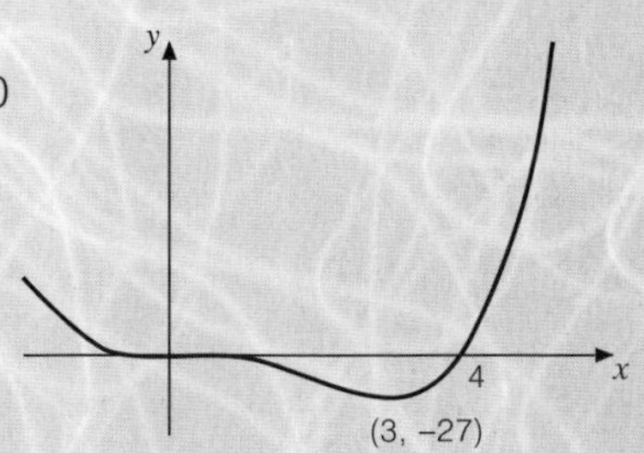

Now consider the following three curves:

$y = x^3$

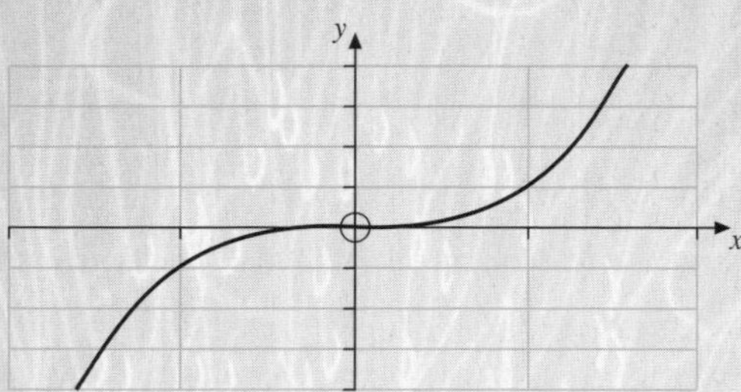

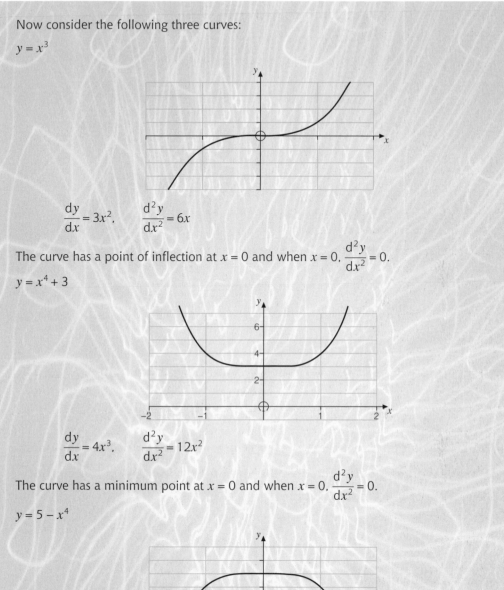

$$\frac{dy}{dx} = 3x^2, \qquad \frac{d^2y}{dx^2} = 6x$$

The curve has a point of inflection at $x = 0$ and when $x = 0$, $\frac{d^2y}{dx^2} = 0$.

$y = x^4 + 3$

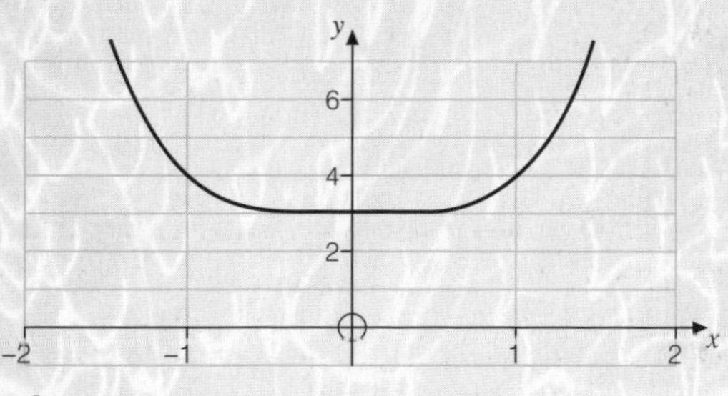

$$\frac{dy}{dx} = 4x^3, \qquad \frac{d^2y}{dx^2} = 12x^2$$

The curve has a minimum point at $x = 0$ and when $x = 0$, $\frac{d^2y}{dx^2} = 0$.

$y = 5 - x^4$

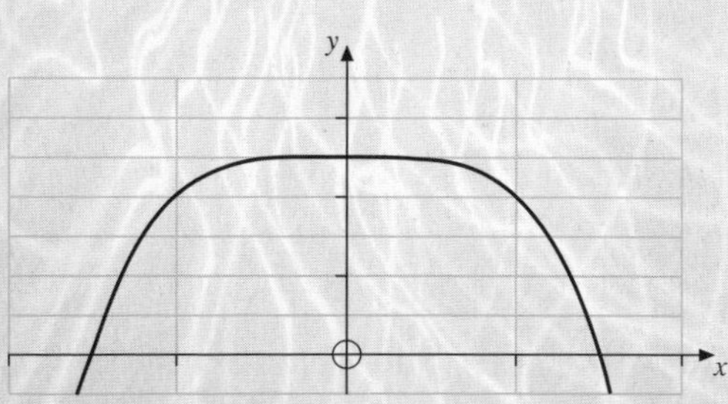

$$\frac{dy}{dx} = -4x^3, \qquad \frac{d^2y}{dx^2} = -12x^2$$

The curve has a maximum point at $x = 0$ and when $x = 0$, $\frac{d^2y}{dx^2} = 0$.

From these three examples we can see that a stationary point where $\frac{d^2y}{dx^2} = 0$ can be a maximum point or a minimum point or a point of inflection.

Having studied this chapter you should know how

- to express a rate of change as a derivative
- to discover whether a function is increasing or decreasing by considering its derivative. In particular use the facts that
  if $f'(x) > 0$ then f is increasing
  if $f'(x) < 0$ then f is decreasing
- to find a stationary point by solving the equation $\frac{dy}{dx} = 0$ and be able to classify the stationary point as a maximum or minimum point by considering the sign of $\frac{dy}{dx}$ either side of the stationary point
- to use differentiation to solve optimisation problems
- to find the equation of the tangent and normal to a curve at a point P. In particular, you should know that

  gradient of tangent at P = gradient of curve at P

  $$\text{gradient of normal at P} = \frac{-1}{\text{gradient of tangent at P}}$$

- to find $\frac{d^2y}{dx^2}$ or $f''(x)$ and be able to use it to classify stationary points. In particular, you should know that

  If $\frac{dy}{dx} = 0$ AND $\frac{d^2y}{dx^2} > 0$ at $x = a$ then $x = a$ is a minimum point

  and

  If $\frac{dy}{dx} = 0$ AND $\frac{d^2y}{dx^2} < 0$ at $x = a$ then $x = a$ is a maximum point

  You should also know that if $\frac{d^2y}{dx^2} = 0$ at a stationary point then the nature of the stationary point must be determined by considering the sign of $\frac{dy}{dx}$ either side of the stationary point

REVISION EXERCISE

1 **a)** Given that $y = x^3 - 6x^2 + 9x + 2$, find $\frac{dy}{dx}$.

b) Hence find the co-ordinates of the stationary points on the curve $y = x^3 - 6x^2 + 9x + 2$ and determine whether each point is a maximum or a minimum.

c) The tangent to the curve at $P(a, b)$ is parallel to the tangent to the curve at $Q(-1, -14)$. Show that the distance between P and Q is $6\sqrt{37}$.

(OCR Jan 2001 P1)

2 a) Find the co-ordinates of the stationary points on the curve

$$y = 2x^3 - 3x^2 - 12x - 7.$$

b) Determine whether each stationary point is a maximum or minimum point.

c) It is given that

$$2x^3 - 3x^2 - 12x - 7 \equiv (x + 1)^2(2x - 7).$$

Sketch the curve $y = (x + 1)^2(2x - 7)$.

d) Write down the set of values of the constant k for which the equation

$$2x^3 - 3x^2 - 12x - 7 = k$$

has exactly one real solution.

(OCR May 2002 P1)

3 a) Given that $y = x^2(x - 3)$, find $\frac{dy}{dx}$.

b) Find the equation of the tangent to the curve $y = x^2(x - 3)$ at the point whose x coordinate is 4. Give your answer in the form $y = mx + c$.

c) The tangent to the curve $y = x^2(x - a)$ at the point with co-ordinates $(5, b)$ has gradient 15. Find the values of the constants a and b.

4 a) Find the equation of the normal to the curve $y = x^2$ at the point P(3, 9). Give your answer in the form $ax + by = c$, where a, b and c are integers.

b) This normal meets the curve again at the point Q. Find the co-ordinates of Q.

5 An open box (that is a box with a base but no lid) is to be made with a square base of side x cm, a height of h cm and a volume of 200 cm^3.

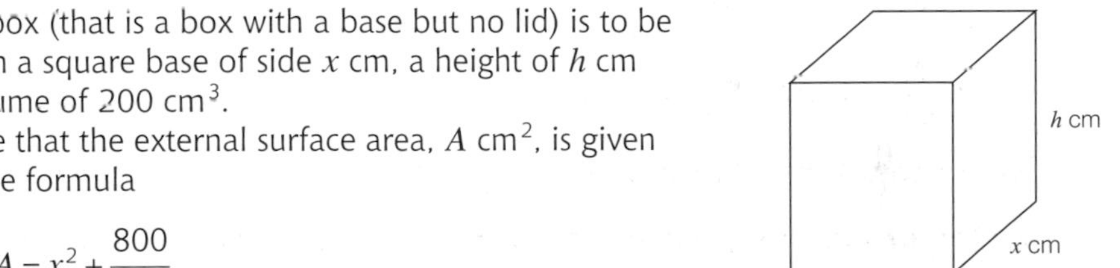

a) Prove that the external surface area, A cm^2, is given by the formula

$$A = x^2 + \frac{800}{x}.$$

b) Find the value of x that should be used to minimise the area of the box.

6 It is given that $g(x) = x^2 + (x^2 - 9.5)^2$.

a) Prove that $g'(x) = 4x^3 - 36x$.

b) Find $g''(x)$.

c) Find the co-ordinates of the stationary points on the curve $y = x^2 + (x^2 - 9.5)^2$.

d) Determine whether each stationary point is a maximum or a minimum point.

Let D be the distance of the point A(0, 9.5) from the point P(x, x^2) on the curve $y = x^2$.

e) Prove that $D^2 = x^2 + (x^2 - 9.5)^2$ and hence find the co-ordinates of the points on the curve $y = x^2$ that are closest to the point A(0, 9.5). Illustrate your answer with a sketch.

7 a) Show that $(x - 2)^2(x + 4) = x^3 - 12x + 16$.

The diagram shows the curve $y = x^3 + 2x - 5$ and the tangent to the curve at the point P(2, 7).

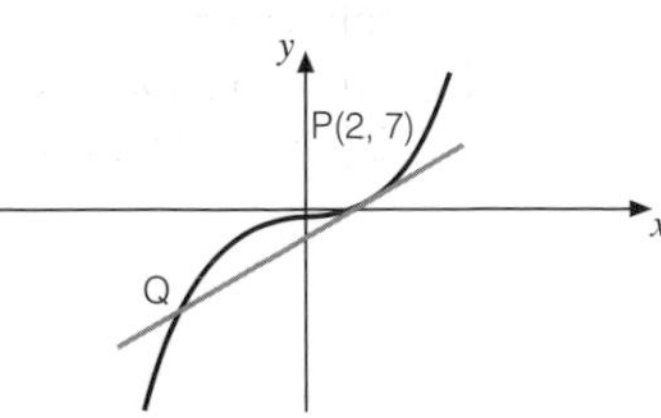

b) Prove that the equation of the tangent is $y = 14x - 21$.

c) Find the co-ordinates of the point Q where the tangent intersects the curve.

8 A sector of a circle is to be made in such a way that the area of the sector is to be 100 cm^2 and the perimeter of the sector is to be as small as possible.

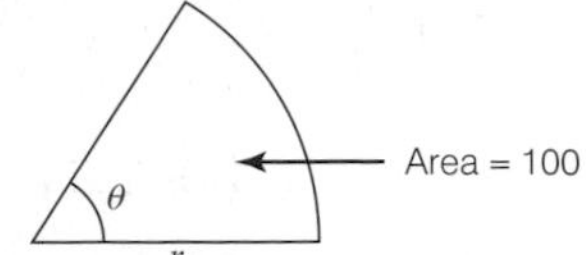

Suppose that the sector has radius r cm and angle $\theta°$.

a) Use the fact that the area of the sector is 100 to write down an equation relating r and θ.

b) Let P denote the perimeter of the sector. Prove that P is given by the formula

$$P = 2r + \frac{200}{r}.$$

c) Find $\frac{\mathrm{d}P}{\mathrm{d}r}$ and hence find the values of r and θ for the sector of area 100 cm^2 with the smallest possible perimeter.

9 The points A and B lie on the curve $y = 2x^{\frac{5}{3}} + x^{\frac{4}{3}}$ and the tangents to the curve at these points are parallel to the line $y = 16x$. Find the x co-ordinates of A and B.
The point A is the point whose co-ordinates are integers. Find the equation of the normal to the curve at A, giving your answer in the form $ax + by + c = 0$.

10 Find the exact value of the x co-ordinate of the stationary point of the curve $y = x - 2(\sqrt[4]{x})^3$. Determine whether the stationary point is a maximum or a minimum.

10 Co-ordinate Geometry: The Circle

The purpose of this chapter is to enable you to

- recognise and use the equation for a circle
- use algebraic methods and geometrical properties of circles to solve problems involving circles and lines

The Equation of a Circle

The diagram shows a circle with centre C(3, 7) and radius 2. The point P(x, y) lies on the circle.

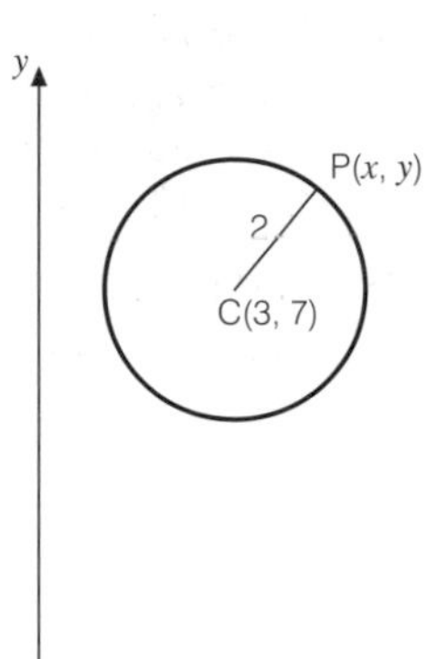

The distance CP is the same as the radius of the circle, so CP = 2.

Using the standard result for distance between two points, you can write

$$\text{CP} = \sqrt{(x-3)^2 + (y-7)^2}.$$

Remember that the distance between (a, b) and (c, d) is given by $\sqrt{(c-a)^2 + (d-b)^2}$.

Combining the two expressions for CP gives

$$\sqrt{(x-3)^2 + (y-7)^2} = 2$$

$$\Rightarrow \quad (x-3)^2 + (y-7)^2 = 4$$

$$\Rightarrow \quad x^2 - 6x + 9 + y^2 - 14y + 49 = 4$$

$$\Rightarrow \quad x^2 + y^2 - 6x - 14y + 54 = 0.$$

This is one way of writing the equation of the circle of radius 2 and centre C(3, 7).

This is a second way of writing the equation of the circle of radius 2 and centre C(3, 7).

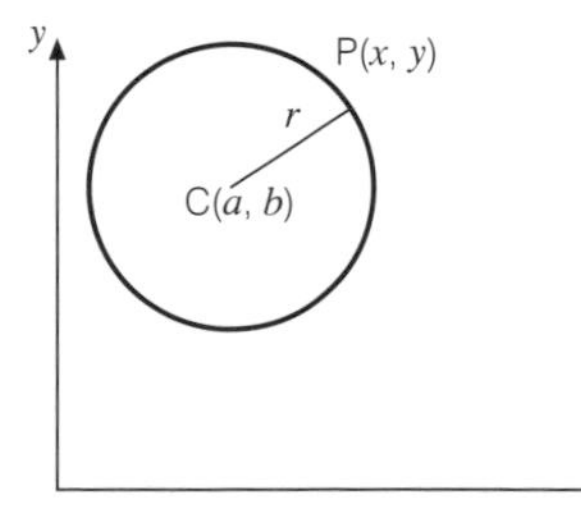

This argument can be generalised to obtain the equation of a circle of radius r and centre C(a, b).

You know that CP is r since it is a radius of the circle and we also know that the distance between the points C(a, b) and P(x, y) is given by $\sqrt{(x-a)^2 + (y-b)^2}$. Combining these results gives

$$\sqrt{(x-a)^2 + (y-b)^2} = r$$

$$\Rightarrow \quad (x-a)^2 + (y-b)^2 = r^2.$$

The equation $(x-a)^2 + (y-b)^2 = r^2$ represents a circle of radius r and centre C(a, b).

You will often be given an equation of the form

$$x^2 + y^2 + 2fx + 2gy + h = 0$$

and be asked to show that it represents a circle. To do this you must use the method of completing the square to rewrite the equation in the form

$$(x-a)^2+(y-b)^2=r^2.$$

The Tangent to a Circle

The equation of a tangent to the circle can be found without any need to do any differentiation.
If P lies on a circle with centre C then recall that the tangent at P is perpendicular to the radius CP.
The gradient of CP can be calculated and the gradient of the tangent can then be deduced using the result for perpendicular lines:

gradient of CP × gradient of tangent = −1.

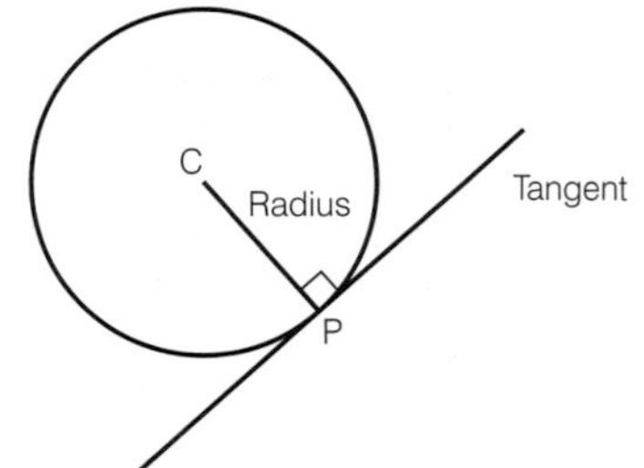

EXAMPLE 1

Show that the equation

$$x^2+y^2+6x-8y-75=0$$

represents a circle and find its centre and radius.
Show that the point A(5, −2) lies on the circle and find the equation of the tangent to the circle at A.

SOLUTION

$$x^2+y^2+6x-8y-75=0$$
$$\Rightarrow \quad x^2+6x+y^2-8y-75=0$$
$$\Rightarrow \quad (x+3)^2-9+(y-4)^2-16-75=0$$
$$\Rightarrow \quad (x+3)^2+(y-4)^2=100$$
$$\Rightarrow \quad (x-(-3))^2+(y-4)^2=10^2$$

Completing the square gives
$x^2+6x=(x+3)^2-9$
and
$y^2-8y=(y-4)^2-16.$

so the equation represents a circle of radius 10 and centre C(−3, 4).

To show that A lies on the circle we could show that the distance AC does equal the radius of the circle:

$$\text{A}(5,-2) \text{ and } \text{C}(-3,4) \Rightarrow \overrightarrow{\mathbf{AC}}=\begin{pmatrix}-8\\6\end{pmatrix} \Rightarrow \text{AC}=\sqrt{(-8)^2+6^2}=10.$$

So A(5, −2) is on the circle with centre C(−3, 4) and radius 10.

To find the equation of the tangent at A:

Alternatively, substitute $x=5$, $y=-2$ into the expression $x^2+y^2+6x-8y-75$ and check the result is 0.

$$\overrightarrow{\mathbf{CA}}=\begin{pmatrix}8\\-6\end{pmatrix}$$

$$\Rightarrow \quad \text{gradient CA}=\frac{y\text{ step}}{x\text{ step}}=\frac{-6}{8}=-\frac{3}{4}$$

$$\Rightarrow \quad \text{gradient of tangent}=\tfrac{4}{3}.$$

Remember that

$$\text{gradient of tangent at A}=\frac{-1}{\text{gradient of radius CA}}.$$

EXAMPLE 1 (continued)

The tangent has gradient $\frac{4}{3}$ and passes through (5, –2) so its equation is

$$y - (-2) = \tfrac{4}{3}(x - 5)$$
$$\Rightarrow \quad y + 2 = \tfrac{4}{3}x - \tfrac{20}{3}$$
$$\Rightarrow \quad y = \tfrac{4}{3}x - \tfrac{26}{3} \quad \text{or} \quad 4x - 3y = 26.$$

Remember that the line with gradient m through the point (x_1, y_1) has equation $y - y_1 = m(x - x_1)$.

EXERCISE 1

1 Find the equation of the circles with

a) centre (4, 2) and radius 5
b) centre (6, –4) and radius 7
c) centre (3, 0) and radius 6
d) centre (–2, –3) and radius 6

2 A circle has the points A(2, –1) and B(–4, 7) at opposite ends of a diameter of the circle.

a) Find the length of AB.
b) Find the centre of the circle.
c) Find the equation of the circle.

3 Find the equation of the circle which has the points (8, 16) and (–16, 6) at opposite ends of a diameter.

4 Find the equation of the circle which

a) has centre (4, 1) and passes through the point (1, 5)
b) has centre (–3, 8) and passes through the point (12, 0).

5 Find the centre and radius of the following circles:

a) $(x - 8)^2 + (y - 3)^2 = 7^2$
b) $(x - 3)^2 + (y - 2)^2 = 16$
c) $(x + 5)^2 + (y - 2)^2 = 4$
d) $(x + 4)^2 + (y + 1)^2 = 7$

6 Consider the circle $(x + 2)^2 + (y - 3)^2 = 25$.

a) What is the centre, C, and radius, r, of this circle?
b) Explain why you can be sure that the point P(2, 6) lies on the circle.
c) Find the gradient of the line CP.
d) Write down the gradient of the tangent to the circle at P.
e) Find the equation of the tangent to the circle at P.

7 Show that the point P(4, 2) lies on the circle $(x - 5)^2 + (y - 3)^2 = 2$.
Find the equation of the tangent to this circle at P.

8 **a)** Find values p and q so that $x^2 + 12x = (x + p)^2 + q$.
b) Find the completed square form of $y^2 - 8y$.
c) Hence, or otherwise, find the centre and radius of the circle whose equation is

$$x^2 + y^2 + 12x - 8y + 48 = 0.$$

9 Show that the equation

$$x^2 + 18x + y^2 + 10y + 6 = 0$$

represents a circle and find the centre and radius of this circle.

10 Show that the equation

$$x^2 - 8x + y^2 - 4y + 11 = 0$$

represents a circle and find the centre and radius of this circle.

11 Show that the equation

$$x^2 + y^2 + 2fx + 2gy + c = 0$$

represents a circle if $f^2 + g^2 - c > 0$ and state the centre and radius of the circle.
What does the equation represent if $f^2 + g^2 - c = 0$?
What does the equation represent if $f^2 + g^2 - c < 0$?

Geometrical Properties of Circles

In your earlier studies you will have discovered and used several important geometrical properties of circles. For this module you need to remember these two results:

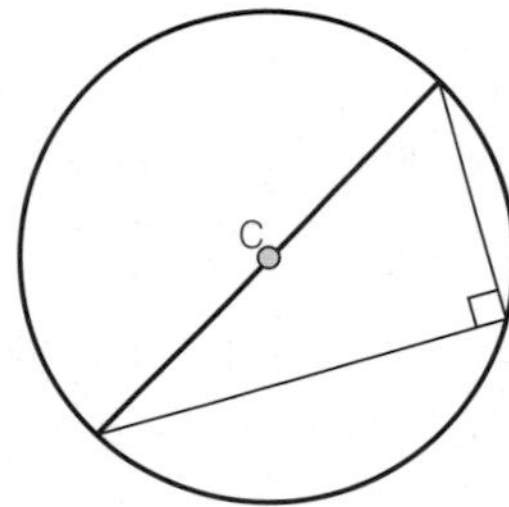

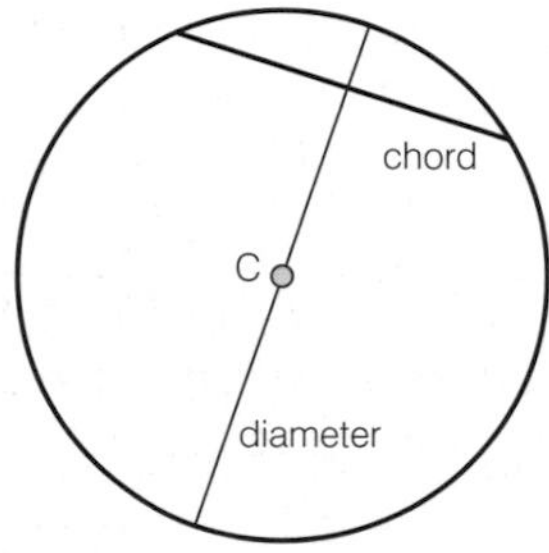

1 The angle in a semi-circle is a right angle.

2 The perpendicular bisector of a chord of a circle is a diameter of the circle.
This also means that the line through the centre which is perpendicular to the chord bisects the chord.

EXAMPLE 2

Find the equation of the circle passing through the points P(4, 6), Q(6, 2) and R(5, −1).

y
P(4, 6)
Q(6, 2)
x
R(5, −1)

SOLUTION

A sketch diagram is usually a good start to a complicated question.

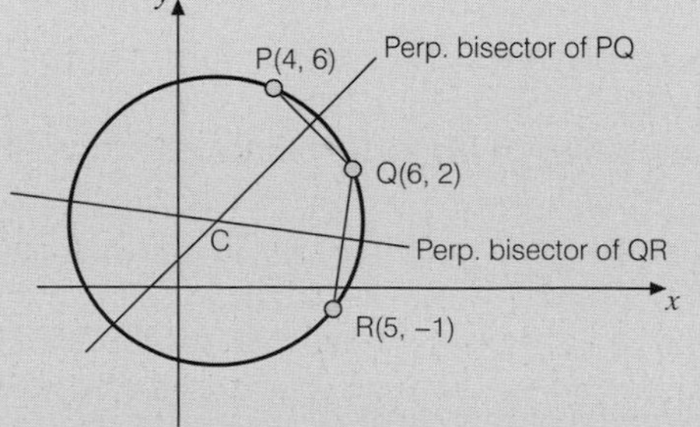

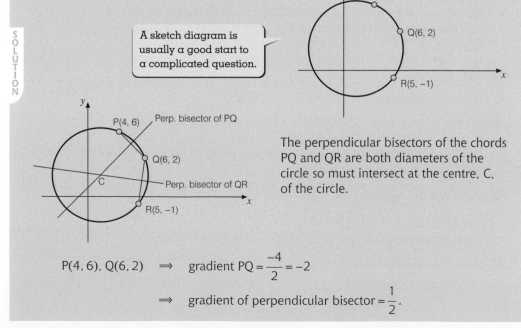

The perpendicular bisectors of the chords PQ and QR are both diameters of the circle so must intersect at the centre, C, of the circle.

$$\text{P}(4, 6), \text{Q}(6, 2) \quad \Rightarrow \quad \text{gradient PQ} = \frac{-4}{2} = -2$$

$$\Rightarrow \quad \text{gradient of perpendicular bisector} = \frac{1}{2}.$$

EXAMPLE 2 (continued)

The midpoint of the chord PQ is the point (5, 4). — $\left(\frac{4+6}{2}, \frac{6+2}{2}\right)$

Since the perpendicular bisector of the chord PQ has gradient $\frac{1}{2}$ and passes through (5, 4), it has equation

$$y - 4 = \tfrac{1}{2}(x - 5)$$
$$\Rightarrow \quad y - 4 = \tfrac{1}{2}x - \tfrac{5}{2}$$
$$\Rightarrow \quad y = \tfrac{1}{2}x + \tfrac{3}{2}.$$

Similarly,

$$Q(6, 2), R(5, -1) \quad \Rightarrow \quad \text{gradient QR} = \frac{-3}{-1} = 3$$
$$\Rightarrow \quad \text{gradient of perpendicular bisector} = -\frac{1}{3}.$$

The midpoint of the chord QR is $(\frac{11}{2}, \frac{1}{2})$.

Since the perpendicular bisector of the chord QR has gradient $-\frac{1}{3}$ and passes through $(\frac{11}{2}, \frac{1}{2})$, it has equation

$$y - \tfrac{1}{2} = -\tfrac{1}{3}(x - \tfrac{11}{2})$$
$$\Rightarrow \quad y - \tfrac{1}{2} = -\tfrac{1}{3}x + \tfrac{11}{6}$$
$$\Rightarrow \quad y = -\tfrac{1}{3}x + \tfrac{7}{3}.$$

The centre of the circle, C, is the point of intersection of these two lines so may be found by solving the line equations simultaneously:

$$\left.\begin{array}{l} y = -\frac{1}{3}x + \frac{7}{3} \\ y = \frac{1}{2}x + \frac{3}{2} \end{array}\right\} \quad \Rightarrow \quad -\tfrac{1}{3}x + \tfrac{7}{3} = \tfrac{1}{2}x + \tfrac{3}{2}$$
$$\Rightarrow \quad \tfrac{5}{6} = \tfrac{5}{6}x$$
$$\Rightarrow \quad x = 1, y = 2.$$

The centre of the circle is therefore the point C(1, 2).

Radius of the circle = CP = $\sqrt{3^2 + 4^2} = 5$.

The circle therefore has equation $(x - 1)^2 + (y - 2)^2 = 25$.

EXAMPLE 3

The line joining the origin to the point A(24, 18) is a diameter of a circle C.
Find its equation.
Two smaller circles, inside the circle C, have their centres on the line OA. The radii of these circles are in the ratio 2 : 1 and they touch each other externally at the point P. The larger circle touches the original circle at O and the smaller circle touches the original circle at A. Find the equations of these two smaller circles and find also the equation of their common tangent at P.

EXAMPLE 3 (continued)

SOLUTION

For the circle C:

$$\text{diameter} = OA = \sqrt{24^2 + 18^2} = 30 \quad \Rightarrow \quad r = 15$$

and

$$\text{centre} = \text{midpoint of OA} = \left(\frac{0+24}{2}, \frac{0+18}{2}\right) = (12, 9).$$

The circle C therefore has equation $(x-12)^2 + (y-9)^2 = 15^2$.

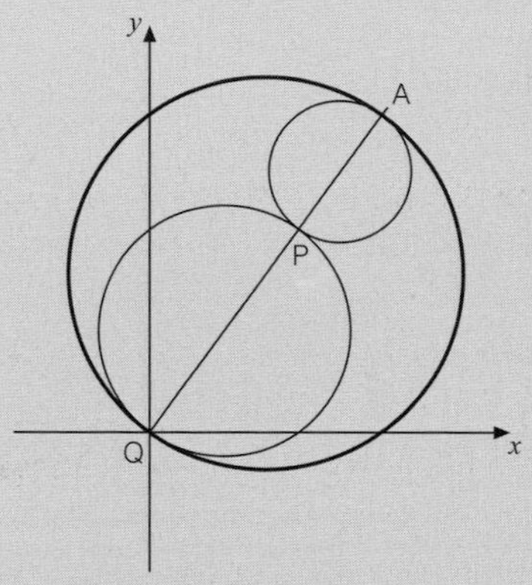

Since the ratio of the radii (or diameters) of the two new circles is 2 : 1, OP must equal 2AP:

$$\vec{\mathbf{OP}} = \frac{2}{3}\vec{\mathbf{OA}} = \frac{2}{3}\begin{pmatrix}24\\18\end{pmatrix} = \begin{pmatrix}16\\12\end{pmatrix} \quad \Rightarrow \quad P(16, 12).$$

Centre of larger circle = midpoint of OP = (8, 6).

Radius of larger circle = $\frac{1}{2}OP = \frac{1}{2}\sqrt{16^2 + 12^2} = 10$.

So the equation of the larger circle is $(x-8)^2 + (y-6)^2 = 10^2$.

Centre of smaller circle = midpoint of PA = (20, 15).

Radius of smaller circle = $\frac{1}{2}PA = \frac{1}{2}\sqrt{8^2 + 6^2} = 5$.

So the equation of the smaller circle is $(x-20)^2 + (y-15)^2 = 5^2$.

The tangent at P is perpendicular to the line OPA which has gradient $\frac{3}{4}$.

The tangent therefore has gradient $-\frac{4}{3}$ and passes through P(16, 12), so it has equation

$$y - 12 = -\tfrac{4}{3}(x-16)$$
$$\Rightarrow \quad y - 12 = -\tfrac{4}{3}x + \tfrac{64}{3}$$
$$\Rightarrow \quad y = -\tfrac{4}{3}x + \tfrac{100}{3} \quad \text{or} \quad 3y + 4x = 100.$$

EXERCISE 2

1 Write down the equation of the circle S which has centre (3, –4) and radius 13.
Find the equation of the tangents to this circle at the points (8, 8) and (–9, 1).
Show that these tangents meet at right angles at the point (–4, 13).

2 Find the centre and radius of the circle

$$x^2 + y^2 + 6x - 4y - 12 = 0$$

and show that the point A(1, 5) lies on this circle.
Find the equation of the tangent to the circle at A.
A second circle has centre (–3, 9) and touches the first circle externally at B. Find the equation of this circle and the co-ordinates of B.
Write down the equation of the common tangent at B and find the point of intersection of this tangent with the tangent to the first circle at A.

3 Find the centres and radii of the circles

$$x^2 + y^2 - 8x - 4y - 5 = 0 \quad \text{and} \quad x^2 + y^2 + 10x + 20y + 25 = 0.$$

Deduce that the two circles touch each other. If the line of centres meets the circles in the points P, Q, R (in that order), find the equation of the circle that has PR as diameter.

4 Find the equation of the circle passing through the points K(3, 3), L(7, 1) and M(7, −5).

5 **a)** Find the centre and radius of the circle whose equation is

$$x^2 + y^2 - 6x + 8y + 21 = 0$$

and sketch this circle.

Let A be the point (5, 0) and P be a point on this circle.

b) Find the maximum and minimum possible values of AP.

6 Two circles have the points A(8, 4) and B(2, −4) as centres, their radii are equal, and they touch externally at the point C.

a) Calculate the radius of each of the circles.
b) Find the co-ordinates of C.
c) Find the equation of the common tangent to the circles at C.
d) Verify that the point D(−7, 9) lies on this tangent.
e) Prove that the circle with centre D which touches each of the two original circles externally has equation

$$x^2 + 14x + y^2 - 18y = 145 - 50\sqrt{10}.$$

Intersection of a Line with a Circle

In chapter 8, you saw how to find the solution of a pair of simultaneous equations where one equation was a quadratic expression and the other was a linear expression. The key steps were to use the linear equation to express one variable in terms of the other and then substitute for this variable in the quadratic expression.
Using this approach, you can find the points of intersection of a line and a circle.

EXAMPLE 4

Find the points of intersection of the line $2y + 3x = 29$ with the circle

$$(x - 1)^2 + y^2 = 65.$$

SOLUTION

The points of intersection are the solutions of the simultaneous equations

$$\left.\begin{aligned} 2y + 3x &= 2 \\ (x-1)^2 + y^2 &= 65 \end{aligned}\right\}.$$

Use the equation $2y + 3x = 29$ to express y in terms of x:

$$\begin{aligned} 2y + 3x &= 29 \\ \Rightarrow \quad 2y &= 29 - 3x \\ \Rightarrow \quad y &= \frac{29 - 3x}{2}. \end{aligned}$$

EXAMPLE 4 (continued)

Substituting this into the second equation, $(x-1)^2 + y^2 = 65$, gives

$$(x-1)^2 + \left(\frac{29-3x}{2}\right)^2 = 65$$

$$\Rightarrow \quad x^2 - 2x + 1 + \frac{(29-3x)^2}{4} = 65$$

$$\Rightarrow \quad 4x^2 - 8x + 4 + (29-3x)^2 = 260$$

$$\Rightarrow \quad 4x^2 - 8x + 4 + 841 - 174x + 9x^2 = 260$$

$$\Rightarrow \quad 13x^2 - 182x + 585 = 0$$

$$[\div 13] \quad \Rightarrow \quad x^2 - 14x + 45 = 0$$

$$\Rightarrow \quad (x-5)(x-9) = 0$$

$$\Rightarrow \quad x - 5 = 0 \quad \text{or} \quad x - 9 = 0$$

$$\Rightarrow \quad x = 5 \quad \text{or} \quad x = 9.$$

Using $y = \dfrac{29-3x}{2}$ gives $x = 5 \Rightarrow y = 7$

and $x = 9 \Rightarrow y = 1$.

So the line $2y + 3x = 29$ intersects the circle $(x-1)^2 + y^2 = 65$ at the points (5, 7) and (9, 1).

The diagram shows the different ways in which a straight line may intersect with a circle.

It can be seen that we might have 0, 1 or 2 points of intersection and, **if there is just one point of intersection then the line is a tangent to the circle.**

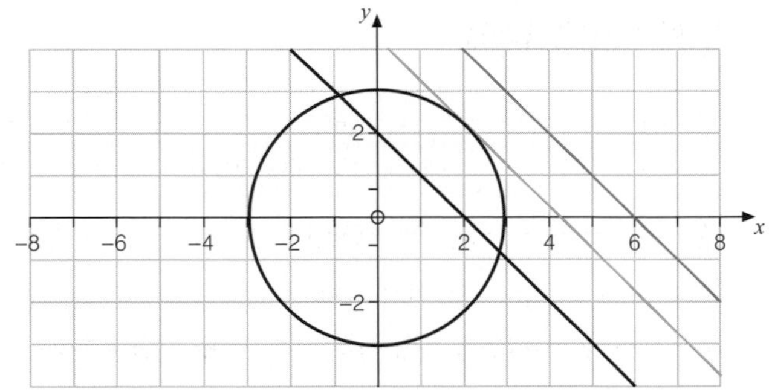

EXAMPLE 5

For what values of the constant c does the line $y = 3x + c$ meet the circle $x^2 + y^2 = 25$ in two distinct points?

SOLUTION

You want the simultaneous equations $\left.\begin{array}{r} y = 3x + c \\ x^2 + y^2 = 25 \end{array}\right\}$ to have two solutions.

Substituting for y in the second equation, you require the equation

$$x^2 + (3x + c)^2 = 25$$

to have two solutions

$\Rightarrow \quad x^2 + 9x^2 + 6cx + c^2 = 25$ must have two solutions

$\Rightarrow \quad 10x^2 + 6cx + (c^2 - 25) = 0$ must have two solutions.

EXAMPLE 5 (continued)

For the quadratic equation $ax^2 + bx + c = 0$ to have two distinct solutions you require $b^2 - 4ac > 0$

$$\Rightarrow \quad (6c)^2 - 4 \times 10 \times (c^2 - 25) > 0$$
$$\Rightarrow \quad 36c^2 - 40c^2 + 1000 > 0$$
$$\Rightarrow \quad 1000 - 4c^2 > 0$$
$$\Rightarrow \quad 250 - c^2 > 0.$$

A graph of $y = 250 - c^2$ will

- be ∩ shaped
- cross the c axis at $\pm\sqrt{250}$.

$\sqrt{250}$ can be simplified to $5\sqrt{10}$.

From the graph

$$250 - c^2 > 0 \quad \Rightarrow \quad -5\sqrt{10} < c < 5\sqrt{10}$$

so

the line $y = 3x + c$ meets the circle $x^2 + y^2 = 25$ in two places provided $-5\sqrt{10} < c < 5\sqrt{10}$.

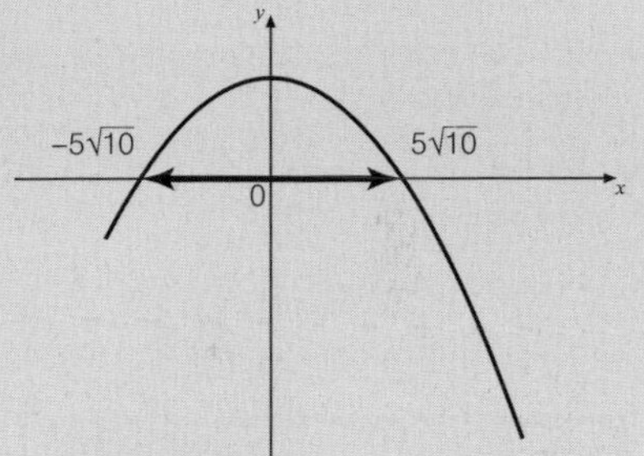

EXERCISE 3

1 **a)** Write down the equation of the circle with centre (0, 0) and radius 10.
b) Find the points of intersection of this circle with the line $y + 7x = 50$.

2 Find the points of intersection of the line $y = 2x - 15$ with the circle

$$(x - 3)^2 + (y - 1)^2 = 25.$$

3 Find the points of intersection of the line $x + 2y = 6$ with the circle

$$x^2 + (y + 2)^2 = 25.$$

4 **a)** Prove that the line $y = 2x - 6$ does not meet the circle $x^2 + y^2 = 4$.
b) Find the exact values of the constant k for which the line $y = 2x - k$ is a tangent to the circle $x^2 + y^2 = 4$.

5 **a)** Write down the equation of a circle of radius 2 and centre the origin.
b) Write down the equation of the straight line through the point (0, –4) which has gradient m.
c) Prove that the x values of the points of intersection of this circle and this line satisfy the equation

$$(1 + m^2)x^2 - 8mx + 12 = 0.$$

d) By considering the discriminant of this equation, describe the relationship between the line and the circle in each of the cases

i) $m = 2$ **ii)** $m = \sqrt{3}$ **iii)** $m = 1$

and illustrate your answer with a sketch diagram in each case.

6 Write down the equation of a circle of radius r and centre (1, 2).
Prove that if $r < \frac{1}{\sqrt{5}}$ then the line $y = 2x - 1$ does not intersect the circle.

7 Find where the line $3x + 4y = 25$ meets the circle $x^2 + y^2 = 25$.
Interpret your answer geometrically.

Having studied this chapter you should know how

- the circle with centre (a, b) and radius r has equation $(x - a)^2 + (y - b)^2 = r^2$
- the equation $x^2 + y^2 + 2fx + 2gy + c = 0$ may represent a circle whose centre and radius may be found through the method of completing the square
- to use simple geometrical properties of circles to answer questions about circles and lines
- to find the points of intersection of a line and a circle

REVISION EXERCISE

1 The ends of a diameter of a circle are at (–2, 4) and (6, –10).
a) Find the centre and radius of the circle.
b) Find the equation of the circle, giving your answer in expanded form (i.e. not involving brackets).
c) Find the equation of the tangent to the circle at the point (–2, 4).

2 The equation of a circle is

$$x^2 + y^2 - 2x + 6y - 15 = 0.$$

a) Find the centre and radius of the circle.
b) The line $y = x + 1$ intersects the circle at A and B. Find the exact length of AB.

3 A circle has equation $(x - 3)^2 + (y + 2)^2 = 9$.
a) Prove that the line $2x - 3y = 12$ passes through the centre of the circle.
b) Find the exact values of the x co-ordinates of the points where the circle crosses the x axis.

4 The circle C has equation $(x + 6)^2 + (y - 8)^2 = r^2$, where r is a positive constant and the line L has equation $y = -3$.
Given that the circle C passes through the origin:
a) find the value of r
b) show that L and C have no points of intersection
c) find the co-ordinates of the points on the line L that are at a distance $5\sqrt{5}$ units from the centre of C.

5 **a)** Find the equation of the perpendicular bisector of A(6, 12) and B(8, 8).

A circle has centre on the x axis and passes through the points A and B.
b) Find the centre and radius of the circle.

6 A circle has the equation $(x - 7)^2 + (y - 7)^2 = 100$.
The circle intersects the x axis at P and Q and intersects the y axis at R and S.
Prove that the quadrilateral PRQS has area 102 square units.

7 The points A, B and C have co-ordinates (3, 1), (5, 2) and (3, 6), respectively.

a) Prove that AB and BC are perpendicular.

b) Hence, or otherwise, find the equation of the circle passing through the points A, B and C.

8 The circles C_1 and C_2 have equations

$$x^2 + y^2 = 20 \quad \text{and} \quad x^2 + y^2 - 24x - 12y + 100 = 0$$

respectively.

a) Find the centre and radius of each circle.

b) Prove that the circles touch externally and find the point where they touch.

c) Prove that the common tangent to the two circles has equation $y + 2x = 10$.

9 **a)** Write down the equation of the circle, C, with centre (3, 0) which passes through the point (0, 0).

b) Write down the equation of the line, L, with gradient m which passes through the point (–2, 0).

c) Show that the x co-ordinates of the points of intersection of L and C are roots of the equation

$$(m^2 + 1)x^2 + (4m^2 - 6)x + 4m^2 = 0.$$

d) Hence find the values of m for which L is a tangent to C.

11 Graphs and Transformations

The purpose of this chapter is to enable you to

- recognise commonly occurring graphs
- sketch a graph from its rule given in factorised form
- understand the effect of simple transformations on graphs

A Catalogue of Commonly Occurring Graphs

During the course of this module and your earlier studies you will have encountered a wide variety of functions and their graphs. Some of the most commonly occurring functions and their associated graphs are shown below.
You should try to ensure both that you can sketch quickly the graph of a function from its equation and that, given a graph, you can suggest a rule for the function.

Function	Graph	Properties
$y = kx$ $(k > 0)$	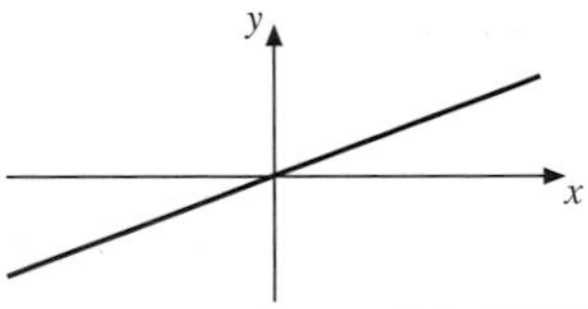	y is proportional to x. As x doubles so does y. As x trebles so does y.
$y = kx^2$ $(k > 0)$	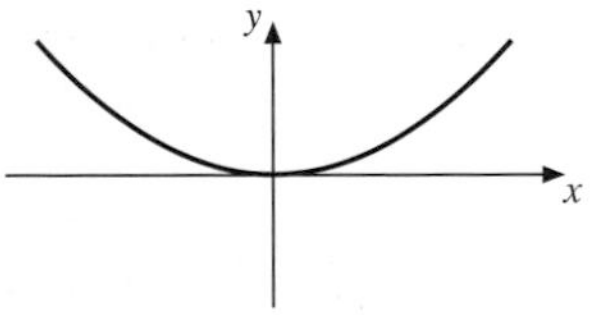	y is proportional to the square of x. As x doubles, y quadruples. As x trebles, y increases by a factor of 9.
$y = kx^3$ $(k > 0)$	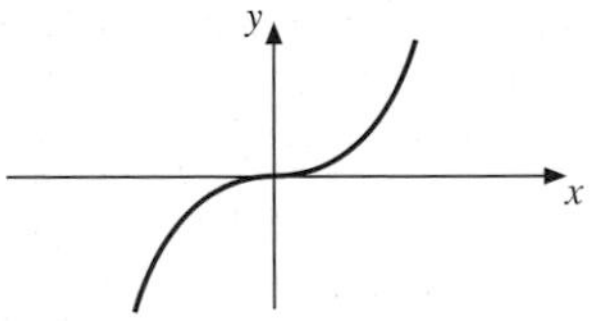	y is proportional to the cube of x. As x doubles, y increases by a factor of 8. As x trebles, y increases by a factor of 27.
$y = kx^n$ n is a positive **even** integer. $(k > 0)$	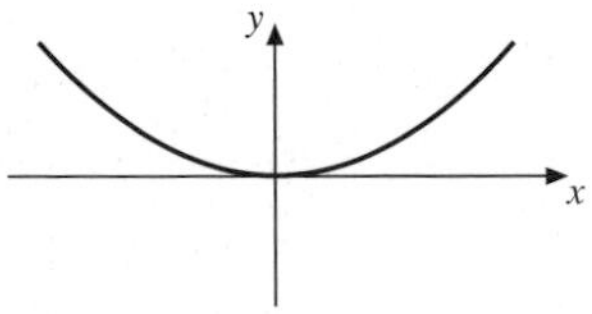	
$y = kx^n$ n is a positive **odd** integer greater than 1. $(k > 0)$	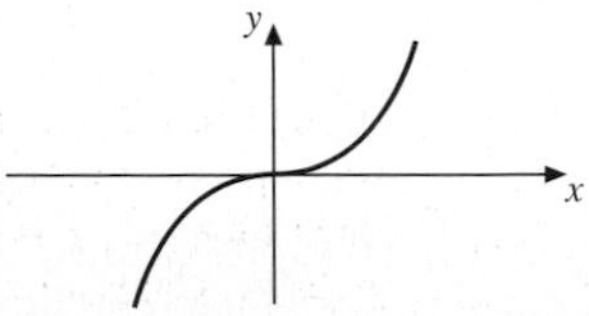	

Function	Graph	Properties
$y = \frac{k}{x}$ $(k > 0)$	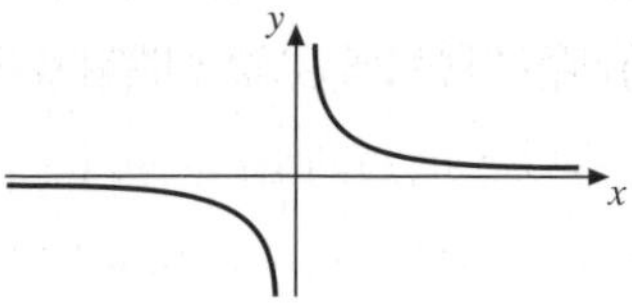	y is inversely proportional to x. As x doubles, y is halved.
$y = \frac{k}{x^2}$ $(k > 0)$	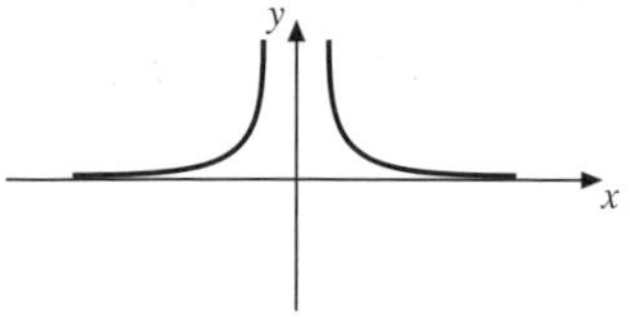	y is inversely proportional to the square of x. As x doubles, the value of y is quartered.
$y = kx^n$ n is a negative **odd** integer. $(k > 0)$	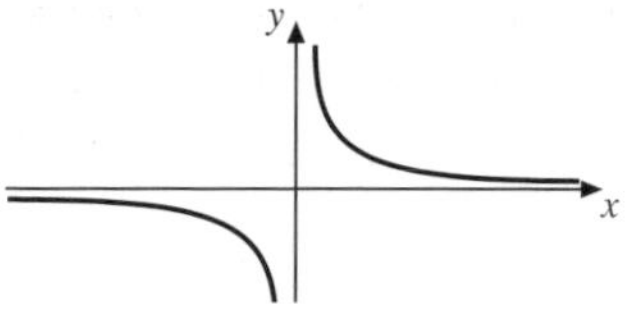	
$y = kx^n$ n is a negative **even** integer. $(k > 0)$	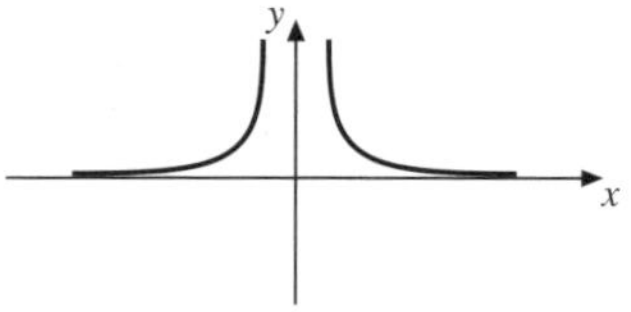	
$y = k\sqrt{x}$ $(k > 0)$	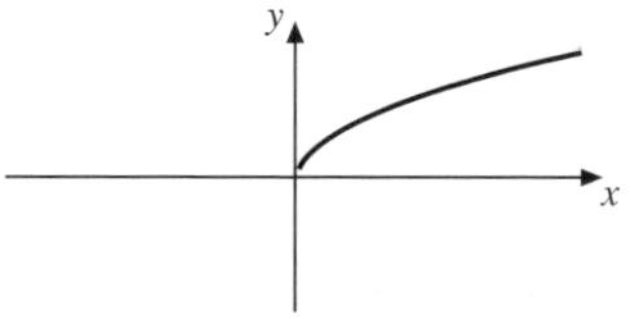	y is proportional to the square root of x. As x increases by a factor of 4, y increases by a factor of 2.

EXAMPLE 1

Suggest a possible equation for the curve shown in the diagram.

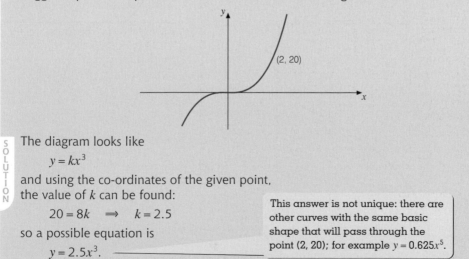

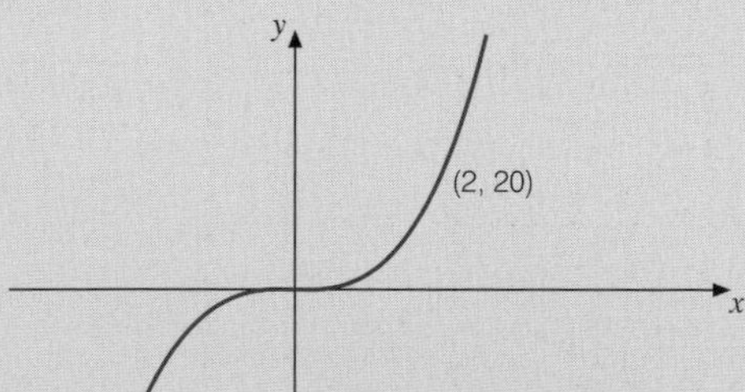

SOLUTION

The diagram looks like

$$y = kx^3$$

and using the co-ordinates of the given point, the value of k can be found:

$$20 = 8k \quad \Rightarrow \quad k = 2.5$$

so a possible equation is

$$y = 2.5x^3.$$

This answer is not unique: there are other curves with the same basic shape that will pass through the point (2, 20); for example $y = 0.625x^5$.

EXAMPLE 2

Give a possible equation for the curve shown in the diagram.

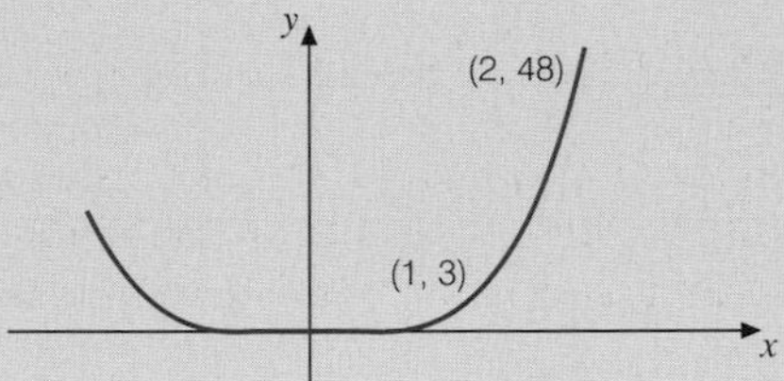

SOLUTION

The curve looks like $y = kx^n$ where n is a positive even number.

Putting $x = 1$, $y = 3$ gives

$$3 = k \times 1^n \quad \Rightarrow \quad k = 3.$$

Putting $x = 2$, $y = 48$ gives

$$48 = 3 \times 2^n \quad \Rightarrow \quad 2^n = 16 \quad \Rightarrow \quad n = 4.$$

A possible equation for this curve is $y = 3x^4$.

EXERCISE 1

Sketch the graphs of

1 $y = 5x^2$

2 $y = \dfrac{8}{x^3}$

3 $y = 5\sqrt{x}$

4 $y = -2x^3$

5 $y = \dfrac{100}{x^4}$

6 $y = -3x$

Determine possible equations for each of the following graphs:

7

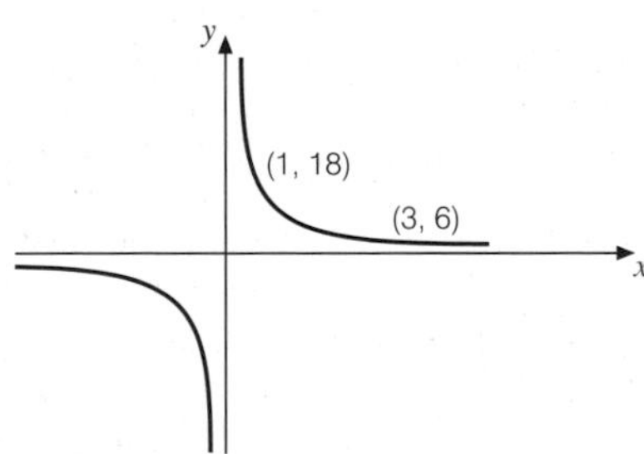

8

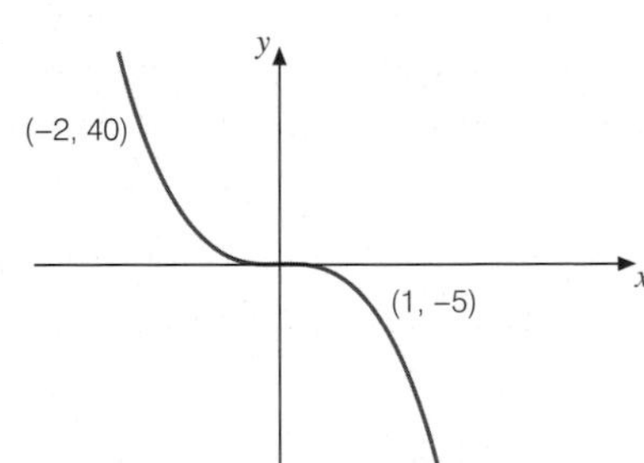

9

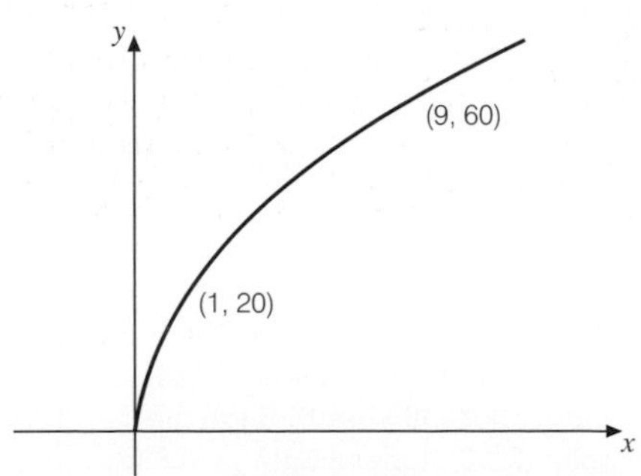

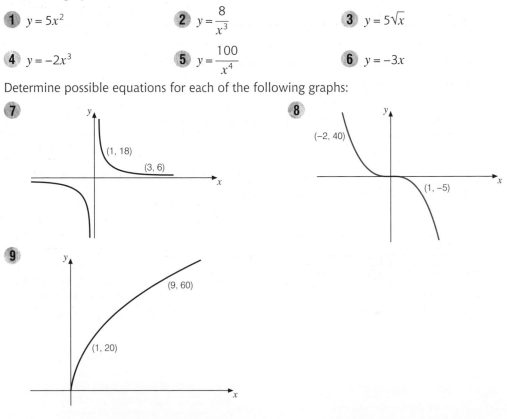

Sketching the Graph of Polynomials Given in Factorised Form

You have already seen in chapter 6 that if a quadratic function is factorisable then a sketch of its graph can readily be drawn.
If you have a polynomial function, $f(x)$, expressed in its factorised form then the roots of the equation $f(x) = 0$ may quickly be found and you can then produce a sketch of the graph of $y = f(x)$ by considering whether $f(x)$ is positive or negative between these roots.

EXAMPLE 3

Sketch the graph of $y = (5 - x)(x + 2)(x - 3)$.

SOLUTION

The value of y is 0 at $x = 5$, -2 or 3.

You can produce a table to show how the sign of y varies.

The top row shows the values of x for which y is 0. These values must be written in numerical value.

This column will show what happens if $x < -2$

This column will show what happens if $-2 < x < 3$

This column will show what happens if $3 < x < 5$

This column will show what happens if $5 < x$

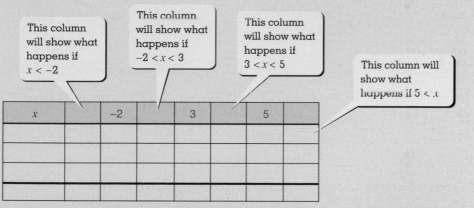

x		−2		3		5	

The next three rows show the values (positive, negative or zero) taken by $(5 - x)$, $(x + 2)$ and $(x - 3)$, respectively.

x		−2		3		5	
$(5 - x)$	+	+	+	+	+	0	−
$(x + 2)$	−	0	+	+	+	+	+
$(x - 3)$	−	−	−	0	+	+	+

This row shows that $5 - x$ is positive if $x < 5$ and negative if $5 < x$

This row shows that $x + 2$ is negative if $x < -2$ and positive if $-2 < x$

This row shows that $x - 3$ is negative if $x < 3$ and positive if $3 < x$

EXAMPLE 3 (continued)

Finally, using the results in each column, the values (positive, negative or zero) of $y = (5 - x)(x + 2)(x - 3)$ are shown.

x		-2		3		5	
$(5 - x)$	+	+	+	+	+	0	–
$(x + 2)$	–	0	+	+	+	+	+
$(x - 3)$	–	–	–	0	+	+	+
$y = (5 - x)(x + 2)(x - 3)$	+	0	–	0	+	0	–

This table, together with the fact that $y = -30$ when $x = 0$, enables us to produce the sketch:

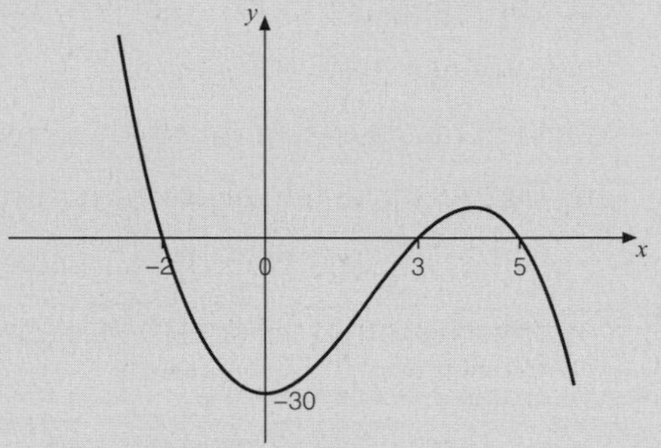

EXAMPLE 4

Sketch the graph of $y = (5 - x)(x + 2)^2$.

SOLUTION

The value of y is clearly 0 at $x = 5$ or -2.

You can now produce a table to show how the sign of y varies:

x		-2		5	
$(5 - x)$	+	+	+	0	–
$(x + 2)^2$	+	0	+	+	+
$y = (5 - x)(x + 2)^2$	+	0	+	0	–

This table, together with the fact that $y = 20$ when $x = 0$, enables you to produce the sketch:

Notice that the curve equation has a repeated factor of $(x + 2)$ and observe that, on the graph, the x axis is a tangent to the curve at $x = -2$.

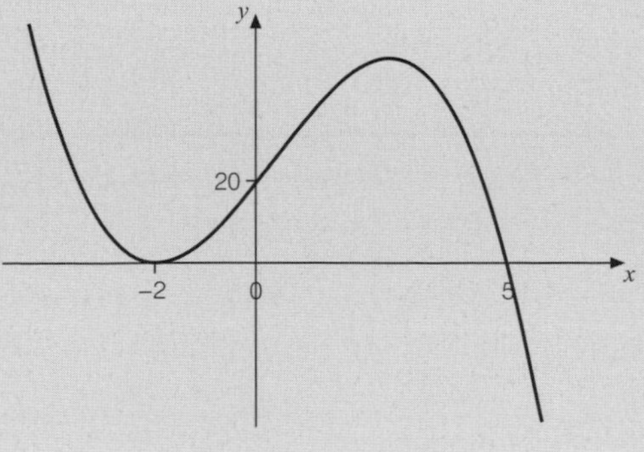

EXAMPLE 5

Given that $x^3 + 5x^2 - 12x - 36$ has a factor of $(x + 2)$, factorise the expression completely and hence sketch the graph of

$$y = x^3 + 5x^2 - 12x - 36.$$

Use your graph to solve the inequality $x^3 + 5x^2 \leqslant 12x + 36$.

SOLUTION

You are told that $x + 2$ is a factor of $x^3 + 5x^2 - 12x - 36$, therefore you can write

$$x^3 + 5x^2 - 12x - 36 = (x + 2)(x^2 + bx + c).$$

The values of b and c can be found by multiplying out the right hand side:

$$\begin{aligned} x^3 + 5x^2 - 12x - 36 &= x^3 + bx^2 + cx + 2x^2 + 2bx + 2c \\ &= x^3 + (2 + b)x^2 + (2b + c)x + 2c. \end{aligned}$$

Looking at the constant term gives

$$-36 = 2c \quad \Rightarrow \quad c = -18$$

and looking at the x^2 term gives

$$5 = 2 + b \quad \Rightarrow \quad b = 3.$$

Now check that the x term works as well: if $b = 3$ and $c = -18$ then $2b + c = -12$ as required!

You now have $x^3 + 5x^2 - 12x - 36 \equiv (x + 2)(x^2 + 3x - 18) \equiv (x + 2)(x + 6)(x - 3)$.

The graph of $y = x^3 + 5x^2 - 12x - 36 = (x + 2)(x + 6)(x - 3)$ will cross the x axis at $x = -6$ or $x = -2$ or $x = 3$.

The table of values for $y = x^3 + 5x^2 - 12x - 36 = (x + 2)(x + 6)(x - 3)$ is therefore

x		−6		−2		3	
$x + 2$	−	−	−	0	+	+	+
$x + 6$	−	0	+	+	+	+	+
$x - 3$	−	−	−	−	−	0	+
$y = (x + 2)(x + 6)(x - 3)$	−	0	+	0	−	0	+

The table, together with the fact that $y = -36$ when $x = 0$, enables us to sketch the graph:

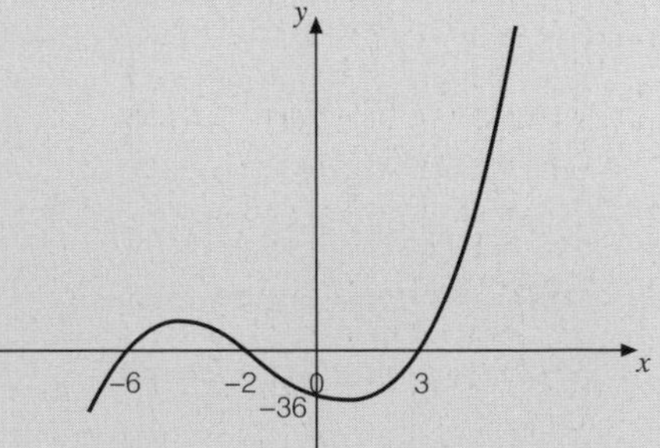

The inequality

$$x^3 + 5x^2 \leqslant 12x + 36$$

can be rewritten as

$$x^3 + 5x^2 - 12x - 36 \leqslant 0.$$

Looking for the places on the graph where the y value is negative gives the solution of this inequality as

$$x \leqslant -6 \quad \text{or} \quad -2 \leqslant x \leqslant 3.$$

EXAMPLE 6

Factorise $f(x) = 2x^4 - 4x^3 - 30x^2$ and hence sketch the graph of $y = f(x)$.

SOLUTION

f has a common factor of $2x^2$, so

$$f(x) = 2x^4 - 4x^3 - 30x = 2x^2(x^2 - 2x - 15) = 2x^2(x + 3)(x - 5).$$

If $y = 2x^2(x + 3)(x - 5)$ then y is 0 when $x = -3$, 0 or 5 and a table can now be produced to show the signs of y:

x		−3		0		53	
$2x^2$	+	+	+	0	+	+	+
$x + 3$	−	0	+	+	+	+	+
$x - 5$	−	−	−	−	−	0	+
$y = 2x^2(x + 3)(x - 5)$	+	0	−	0	−	0	+

and the sketch can now be produced:

Notice that the curve equation has a repeated factor of x and observe that, on the graph, the x axis is a tangent to the curve at $x = 0$.

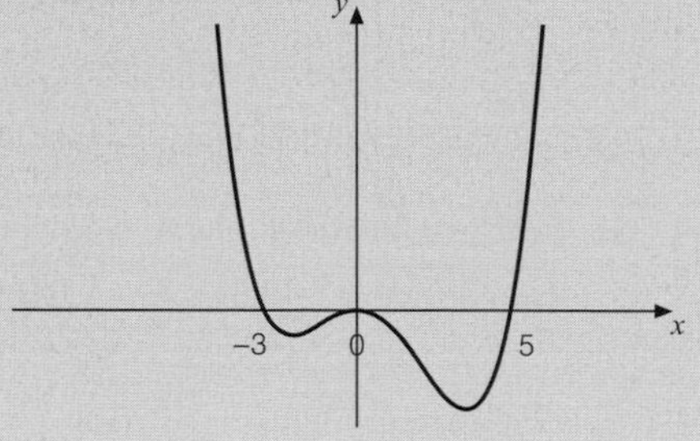

EXAMPLE 7

Find a suitable equation for the curve shown in the diagram below.

SOLUTION

The fact that the curve crosses the x axis at −3 suggests a factor of $(x + 3)$.

The fact that the curve crosses the x axis at 1 suggests a factor of $(x - 1)$.

The fact that the curve crosses the x axis at 4 suggests a factor of $(x - 4)$.

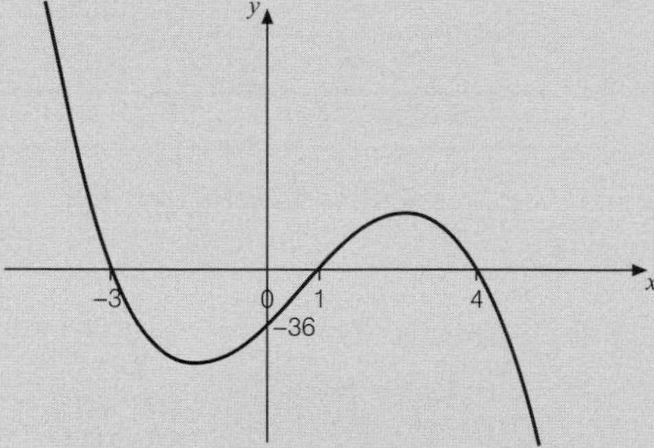

This suggests

$$y = k(x + 3)(x - 1)(x - 4)$$

where k is some constant.

When $x = 0$ you want $y = -36$ so

$$-36 = k \times 3 \times (-1) \times (-4)$$
$$\Rightarrow \quad -36 = 12k$$
$$\Rightarrow \quad k = -3.$$

A possible rule for the graph is $\quad y = -3(x + 3)(x - 1)(x - 4).$

You could use a graphical calculator to check this answer.

EXAMPLE 8

Find a suitable equation for the curve shown in the diagram below.

SOLUTION

The fact that the curve touches the x axis at -2 suggests a factor of $(x+2)^2$.

The fact that the curve crosses the x axis at 4 suggests a factor of $(x-4)$.

This suggests $y = k(x+2)^2(x-4)$.

Putting $x = 0$ gives

$$64 = k \times 4 \times -4 \quad \Rightarrow \quad k = -4.$$

A possible rule for this graph is
$y = -4(x+2)^2(x-4)$.

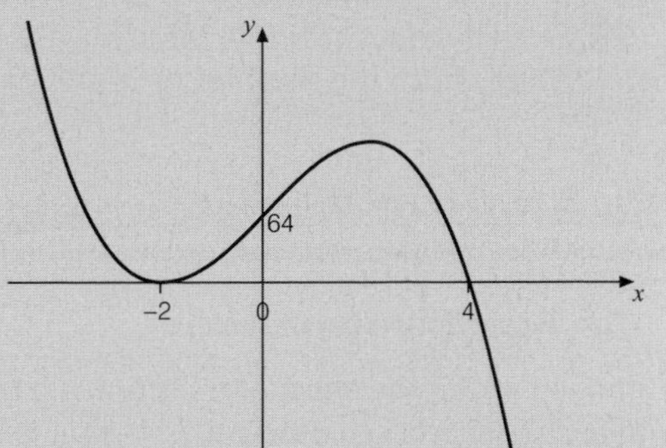

EXERCISE 2

Sketch, without using the methods of calculus or your graphical calculator, the curves whose equations are

1 $y = 3(x-2)(x+6)$

2 $y = (4-x)(x+2)$

3 $y = (x+3)(x-3)(x-1)$

4 $y - (x-3)^2(x+2)$

5 $y = x(3-x)(x+5)$

6 $y = x(3-x)^2$

Find possible equations for each of the following curves:

7

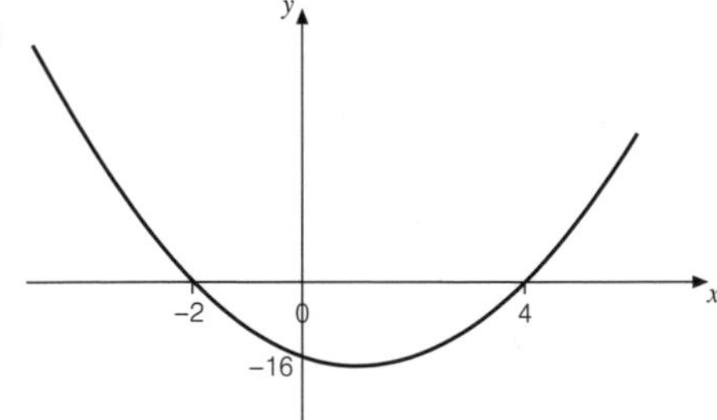

8

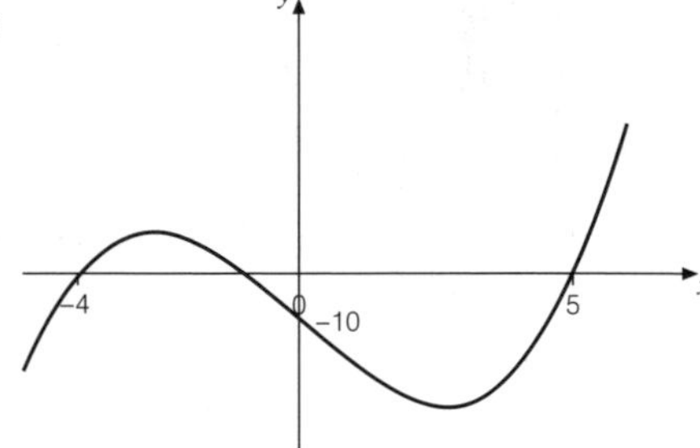

9

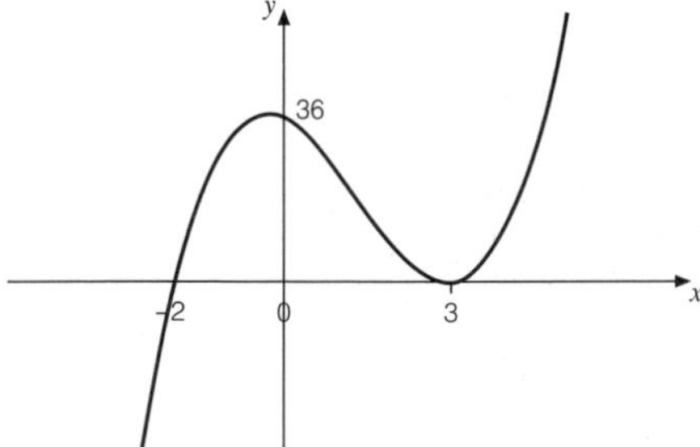

10 Factorise $4x^4 - 24x^3 + 36x^2$ completely and hence sketch the graph of $y = 4x^4 - 24x^3 + 36x^2$.

11 Given that $(x - 5)$ is a factor of $2x^3 - 17x^2 + 41x - 30$, factorise $2x^3 - 17x^2 + 41x - 30$ completely and hence sketch the graph of $y = 2x^3 - 17x^2 + 41x - 30$.
Solve the inequality $2x^3 + 41x \leqslant 17x^2 + 30$.

12 Given that $(x - 2)$ is a factor of $x^3 - 39x + 70$, factorise $x^3 - 39x + 70$ completely and hence sketch the graph of $y = x^3 - 39x + 70$.

The Effect of Transformations on Graphs

The Basic Transformations

In this chapter the effect of transformations upon curves will be investigated, in particular the effect of reflections in the x axis, translations and stretches parallel to either the x axis or the y axis. The diagrams below show the effect of these transformations on a rectangle whose vertices are O(0, 0), A(0, 1), B(2, 1) and C(2, 0).

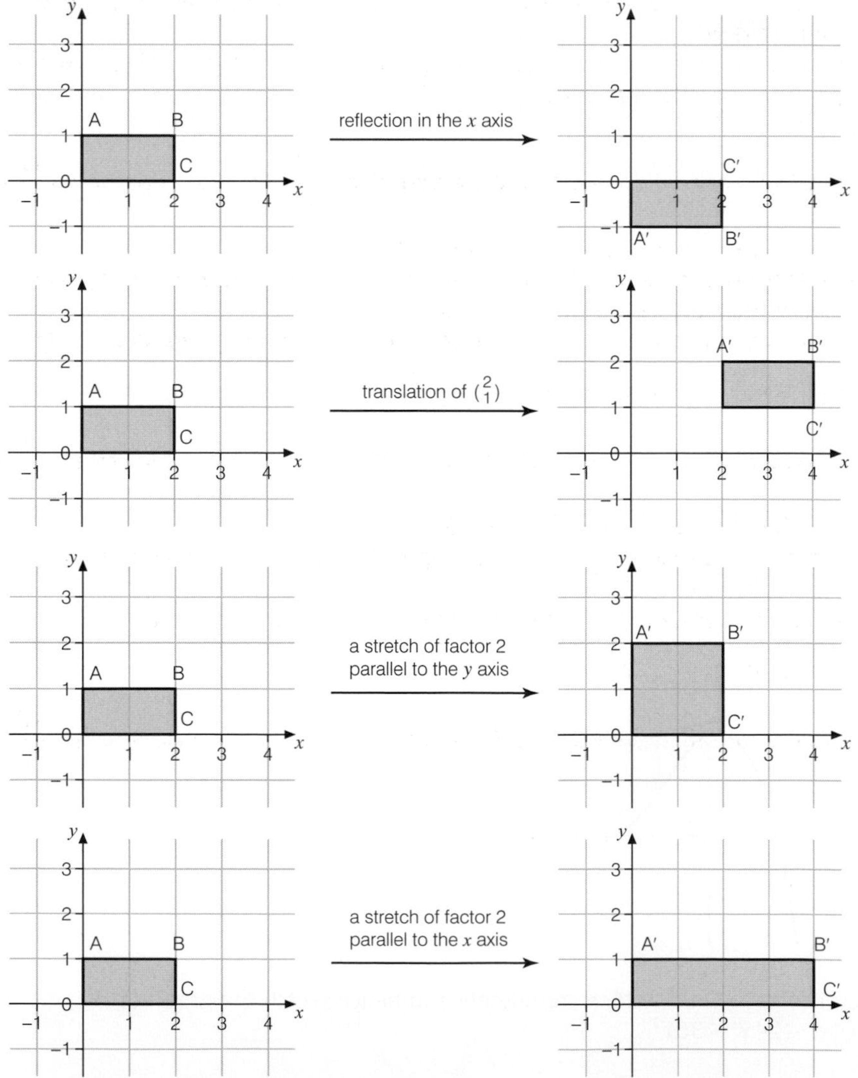

EXERCISE 3

In this exercise the transformations required to map curves onto related curves will be investigated. You should work through this exercise before reading the next section.

Use a graphical calculator or a graph drawing program on a computer to answer this exercise.

1 **a)** Sketch the graphs of $y = x^2$ and $y = -x^2$.
What transformation will map the graph of $y = x^2$ onto the graph of $y = -x^2$?

b) Sketch the graphs of $y = x^3$ and $y = -x^3$.
What transformation will map the graph of $y = x^3$ onto the graph of $y = -x^3$?

c) Sketch the graphs of $y = \frac{12}{x}$ and $y = -\frac{12}{x}$.

What transformation will map the graph of $y = \frac{12}{x}$ onto the graph of $y = -\frac{12}{x}$?

2 **a)** Sketch the graphs of $y = x^2$ and $y = x^2 + 7$.
What transformation will map the graph of $y = x^2$ onto the graph of $y = x^2 + 7$?

b) Sketch the graphs of $y = x^3$ and $y = x^3 - 4$.
What transformation will map the graph of $y = x^3$ onto the graph of $y = x^3 - 4$?

c) Sketch the graphs of $y = \frac{12}{x}$ and $y = \frac{12}{x} + 5$.

What transformation will map the graph of $y = \frac{12}{x}$ onto the graph of $y = \frac{12}{x} + 5$?

3 **a)** Sketch the graphs of $y = x^2$ and $y = (x - 2)^2$.
What transformation will map the graph of $y = x^2$ onto the graph of $y = (x - 2)^2$?

b) Sketch the graphs of $y = x^3$ and $y = (x + 2)^3$.
What transformation will map the graph of $y = x^3$ onto the graph of $y = (x + 2)^3$?

c) Sketch the graphs of $y = \frac{12}{x}$ and $y = \frac{12}{x - 3}$.

What transformation will map the graph of $y = \frac{12}{x}$ onto the graph of $y = \frac{12}{x - 3}$?

4 **a)** Sketch the graphs of $y = x^2$ and $y = 3x^2$.
What transformation will map the graph of $y = x^2$ onto the graph of $y = 3x^2$?

b) Sketch the graphs of $y = x^3$ and $y = 2x^3$.
What transformation will map the graph of $y = x^3$ onto the graph of $y = 2x^3$?

c) Sketch the graphs of $y = \frac{12}{x}$ and $y = \frac{48}{x}$.

What transformation will map the graph of $y = \frac{12}{x}$ onto the graph of $y = \frac{48}{x}$?

The Effect of Reflections and Translations upon Graphs

The diagram shows the graph of $y = x^2$ and its image after a reflection in the x axis.

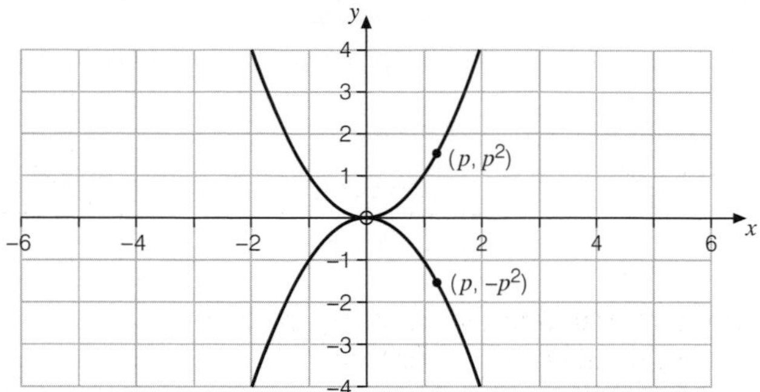

You can see that the point (p, p^2) is mapped to the point $(p, -p^2)$ and it is clear that this new point lies on the curve $y = -x^2$.

Now consider what happens to a point $P(p, f(p))$ on the curve $y = f(x)$ if the curve is reflected in the x axis:

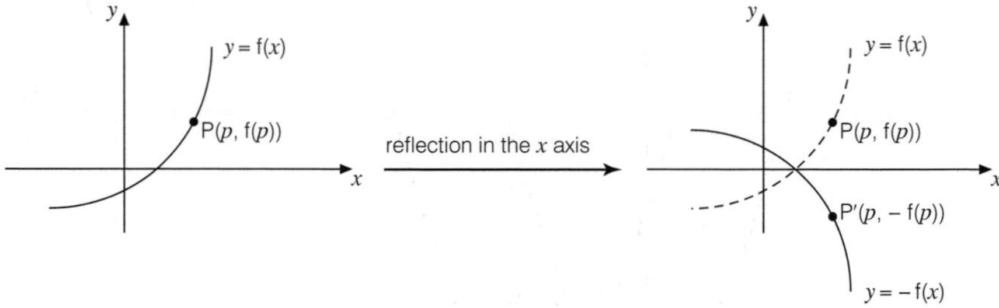

The point $P(p, f(p))$ is reflected to the point $P'(p, -f(p))$.
The point P′ clearly lies on the curve whose equation is $y = -f(x)$.

$$y = f(x) \xrightarrow{\text{reflection in } x \text{ axis}} y = -f(x)$$

The diagram shows the graph of $y = x^2$ and its image after a translation of $\begin{pmatrix} 0 \\ 3 \end{pmatrix}$.

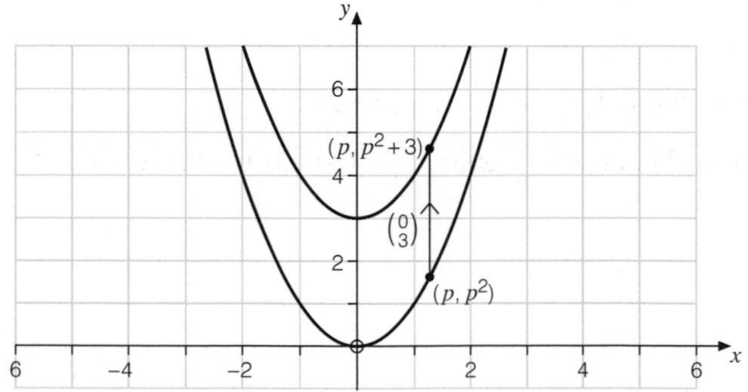

You can see that the point (p, p^2) is mapped to the point $(p, p^2 + 3)$ and it is clear that this point lies on the curve $y = x^2 + 3$.

Now consider what happens to a point $P(p, f(p))$ on the curve $y = f(x)$ if the curve is translated by $\begin{pmatrix} 0 \\ b \end{pmatrix}$:

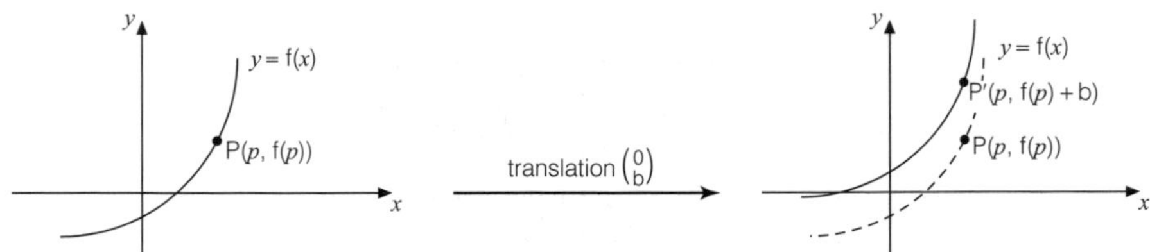

The point $P(p, f(p))$ is translated to the point $P'(p, f(p) + b)$.
The point P′ clearly lies on the curve $y = f(x) + b$.

$$y = f(x) \xrightarrow{\text{translation} \begin{pmatrix} 0 \\ b \end{pmatrix}} y = f(x) + b$$

The diagram shows the graph of $y = x^2$ and its image after a translation of $\begin{pmatrix} 2 \\ 0 \end{pmatrix}$.

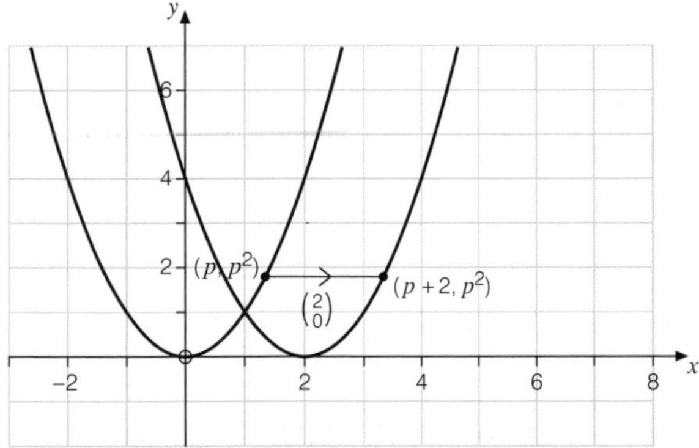

You can see that the point (p, p^2) is mapped to the point $(p + 2, p^2)$.

Now consider this point on the image curve:

$$x = p + 2 \quad \Rightarrow \quad p = x - 2$$

and $\quad y = p^2 \quad \Rightarrow \quad y = (x - 2)^2$.

So the equation of the image curve is $y = (x - 2)^2$.

Now consider what happens to a point $P(p, f(p))$ on the curve $y = f(x)$ if the curve is translated by $\begin{pmatrix} a \\ 0 \end{pmatrix}$:

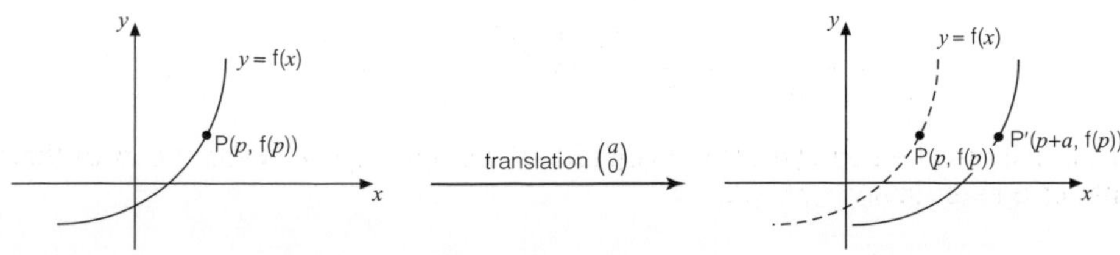

If the curve $y = f(x)$ is translated by $\begin{pmatrix} a \\ 0 \end{pmatrix}$ then the point $P(p, f(p))$ is translated to the point $P'(p + a, f(p))$.

Now consider this point on the image curve:

$$x = p + a \quad \Rightarrow \quad p = x - a$$

and $\quad y = f(p) \quad \Rightarrow \quad y = f(x - a)$.

So the equation of the image curve is $y = f(x - a)$.

$$y = f(x) \xrightarrow{\text{translation} \begin{pmatrix} a \\ 0 \end{pmatrix}} y = f(x - a)$$

EXAMPLE 9

Find the equation of the image of the curve $y = 2x^3$ after a translation of $\begin{pmatrix} -5 \\ 0 \end{pmatrix}$.

SOLUTION

You know that $\quad y = f(x) \xrightarrow{\text{translation} \begin{pmatrix} a \\ 0 \end{pmatrix}} y = f(x - a)$

so $\quad y = 2x^3 \xrightarrow{\text{translation} \begin{pmatrix} -5 \\ 0 \end{pmatrix}} y = 2(x - (-5))^3$.

This means that the image curve is $y = 2(x + 5)^3$.

EXAMPLE 10

Find the equation of the image of the curve $y = x^2 + 5x + 2$

a) after a translation of $\begin{pmatrix} 0 \\ 7 \end{pmatrix}$

b) after a translation of $\begin{pmatrix} 2 \\ 0 \end{pmatrix}$.

SOLUTION

a) You know that $\quad y = f(x) \xrightarrow{\text{translation} \begin{pmatrix} 0 \\ b \end{pmatrix}} y = f(x) + b$

so $\quad y = x^2 + 5x + 2 \xrightarrow{\text{translation} \begin{pmatrix} 0 \\ 7 \end{pmatrix}} y = x^2 + 5x + 2 + 7$.

This means that the image curve is $y = x^2 + 5x + 9$.

b) You know that $\quad y = f(x) \xrightarrow{\text{translation} \begin{pmatrix} a \\ 0 \end{pmatrix}} y = f(x - a)$

so $\quad y = x^2 + 5x + 2 \xrightarrow{\text{translation} \begin{pmatrix} 2 \\ 0 \end{pmatrix}} y = (x - 2)^2 + 5(x - 2) + 2$.

This means that the image curve is $\quad y = (x - 2)^2 + 5(x - 2) + 2$

$$\Rightarrow \quad y = x^2 - 4x + 4 + 5x - 10 + 2$$

$$\Rightarrow \quad y = x^2 + x - 4.$$

EXAMPLE 11

Find the transformation that maps

a) $y = x^2$ onto $y = (x - 9)^2$

b) $y = \frac{1}{x}$ onto $y = \frac{1}{x + 2}$

c) $y = \frac{1}{x}$ onto $y = \frac{1}{x} - 4$.

SOLUTION

Using the three basic results you can write down:

a) $y = x^2 \xrightarrow{\text{translation} \begin{pmatrix} 9 \\ 0 \end{pmatrix}} y = (x - 9)^2$

b) $y = \frac{1}{x} \xrightarrow{\text{translation} \begin{pmatrix} -2 \\ 0 \end{pmatrix}} y = \frac{1}{x - (-2)} = \frac{1}{x + 2}$

c) $y = \frac{1}{x} \xrightarrow{\text{translation} \begin{pmatrix} 0 \\ -4 \end{pmatrix}} y = \frac{1}{x} + (-4) = \frac{1}{x} - 4$

EXERCISE 4

1 Sketch the graph of $y = \frac{12}{x}$. Sketch the image of this graph after a reflection in the x axis. What is the equation of the image curve?

2 Write down the equations of the images of the following curves under the given transformations:

a) $y = x^2$ under the translation $\begin{pmatrix} 0 \\ 4 \end{pmatrix}$

b) $y = \sqrt{x}$ under the translation $\begin{pmatrix} 4 \\ 0 \end{pmatrix}$

c) $y = \frac{1}{x}$ under the translation $\begin{pmatrix} 0 \\ -5 \end{pmatrix}$

d) $y - x^3$ under the translation $\begin{pmatrix} -2 \\ 0 \end{pmatrix}$

e) $y = \sqrt{x}$ under the translation $\begin{pmatrix} 0 \\ 2 \end{pmatrix}$

f) $y = \sqrt{x}$ under the translation $\begin{pmatrix} 5 \\ 0 \end{pmatrix}$

3 Write down the transformations which move

a) $y = x^2$ onto $y = (x + 7)^2$

b) $y = x^2$ onto $y = x^2 + 7$

c) $y = x^6$ onto $y = -x^6$

d) $y = x^2$ onto $y = (x - 3)^2$

e) $y = x^4$ onto $y = x^4 + 10$

f) $y = \frac{8}{x^2}$ onto $y = \frac{-8}{x^2}$

g) $y = x^4$ onto $y = (x - 10)^4$

h) $y = \sqrt{x}$ onto $y = 2 + \sqrt{x - 4}$

4 **a)** Sketch the graph of $y = x^4$.

b) Find the equation of the tangent to the curve $y = x^4$ at the point (1, 1).

c) Describe fully the geometrical transformation that will map the curve $y = x^4$ onto the curve $y = (x - 4)^4$. Hence sketch the graph of $y = (x - 4)^4$.

d) Use your previous answers to deduce the equation of the tangent to the curve $y = (x - 4)^4$ at the point (5, 1).

e) Find also the equation of the tangent to the curve $y = (x - 4)^4$ at the point (3, 1).

The Effect of Stretches on Graphs

The diagram shows the graph of $y = x^2$ and its image after a stretch of scale factor 2 in the y direction:

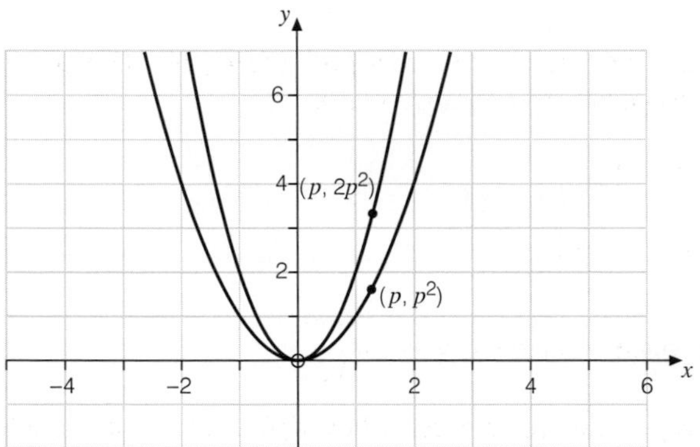

You can see that the point (p, p^2) is mapped to the point $(p, 2p^2)$ and it is clear that this point lies on the curve $y = 2x^2$.

Now consider what happens to a point $P(p, f(p))$ on the curve $y = f(x)$ if the curve is stretched by scale factor k in the y direction:

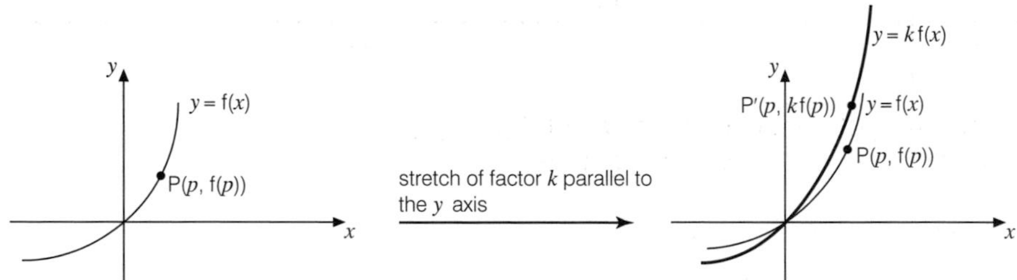

The point $P(p, f(p))$ is mapped to the point $P'(p, kf(p))$.
The point P′ clearly lies on the curve $y = kf(x)$.

$$y = f(x) \xrightarrow{\text{stretch of scale factor } k \text{ in } y \text{ direction}} y = kf(x)$$

The diagram shows the graph of $y = x^2$ and its image after a stretch of scale factor 3 in the x direction:

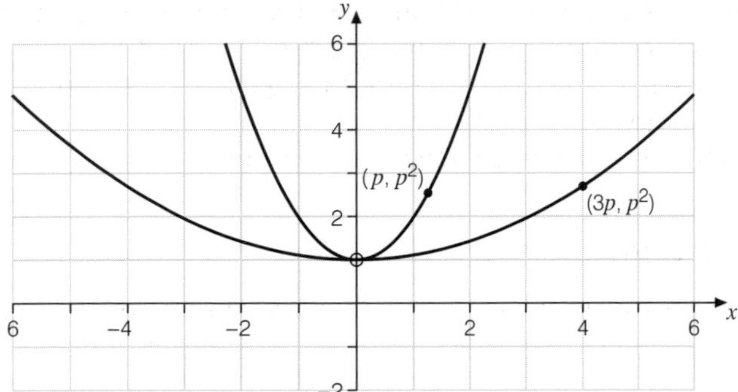

You can see that the point (p, p^2) is mapped to the point $(3p, p^2)$.
Now consider this point on the image curve:

$$x = 3p \quad \Rightarrow \quad p = \frac{x}{3}$$

and $\quad y = p^2 \quad \Rightarrow \quad y = \left(\frac{x}{3}\right)^2.$

So the equation of the image curve is $y = \left(\frac{x}{3}\right)^2$.

Now consider what happens to a point $\mathrm{P}(p, \mathrm{f}(p))$ on the curve $y = \mathrm{f}(x)$ if the curve is stretched by scale factor k in the x direction:

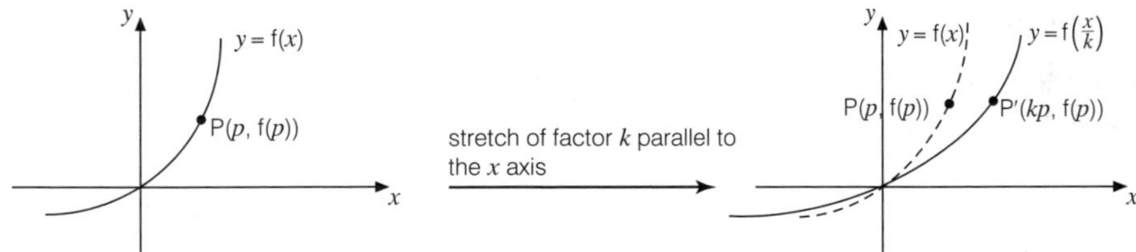

The point $\mathrm{P}(p, \mathrm{f}(p))$ is mapped to the point $\mathrm{P}'(kp, \mathrm{f}(p))$.
Now consider this point on the image curve:

$$x = kp \quad \Rightarrow \quad p = \frac{x}{k}$$

and $\quad y = \mathrm{f}(p) \quad \Rightarrow \quad y = \mathrm{f}\left(\frac{x}{k}\right).$

So the equation of the image curve is $y = \mathrm{f}\left(\frac{x}{k}\right)$.

$$y = \mathrm{f}(x) \xrightarrow{\text{stretch of scale factor } k \text{ in } x \text{ direction}} y = \mathrm{f}\left(\frac{x}{k}\right)$$

This last result could also be written as

$$y = \mathrm{f}(x) \xrightarrow{\text{stretch of scale factor } \frac{1}{k} \text{ in } x \text{ direction}} y = \mathrm{f}(kx)$$

EXAMPLE 12

Find the equation of the image of

a) $y = \dfrac{3}{x}$ after a stretch of scale factor 2 in the y direction;

b) $y = x^2 - 2x$ after a stretch of scale factor 2 in the x direction;

c) $y = (x - 2)(2x + 1)$ after a stretch of scale factor $\dfrac{1}{4}$ in the x direction.

SOLUTION

a) You know that $y = \text{f}(x) \xrightarrow{\text{stretch of scale factor } k \text{ in } x \text{ direction}} y = k\text{f}(x)$

so $y = \dfrac{3}{x} \xrightarrow{\text{stretch of scale factor 2 in } y \text{ direction}} y = 2 \times \dfrac{3}{x}.$

This means that the image curve is $y = \dfrac{6}{x}$.

b) We know that $y = \text{f}(x) \xrightarrow{\text{stretch of scale factor } k \text{ in } x \text{ direction}} y = \text{f}\left(\dfrac{x}{k}\right)$

so $y = x^2 - 2x \xrightarrow{\text{stretch of scale factor 2 in } x \text{ direction}} y = \left(\dfrac{x}{2}\right)^2 - 2\left(\dfrac{x}{2}\right).$

This means that the image curve is $y = \dfrac{x^2}{4} - x$.

c) We know that $y = \text{f}(x) \xrightarrow{\text{stretch of scale factor } \frac{1}{k} \text{ in } x \text{ direction}} y = \text{f}(kx)$

so $y = (x - 2)(2x + 1) \xrightarrow{\text{stretch of scale factor } \frac{1}{4} \text{ in } x \text{ direction}} y = ((4x) - 2)(2(4x) + 1).$

This means that the image curve is $y = (4x - 2)(8x + 1)$.

EXAMPLE 13

If $\text{f}(x) = (x - 4)^2(x + 2)$ draw, on separate diagrams, sketches of the graphs of

a) $y = \text{f}(x)$ **b)** $y = 3\text{f}(x)$ **c)** $y = \text{f}(2x)$

SOLUTION

The roots of the equation $(x - 4)^2(x + 2) = 0$ are 4 and −2 so these are the values that must be shown in the top row of the table:

x		−2		4	
$(x - 4)^2$	+	+	+	0	+
$(x + 2)$	−	0	+	+	+
$y = (x - 4)^2(x + 2)$	−	0	+	0	+

EXAMPLE 13 (continued)

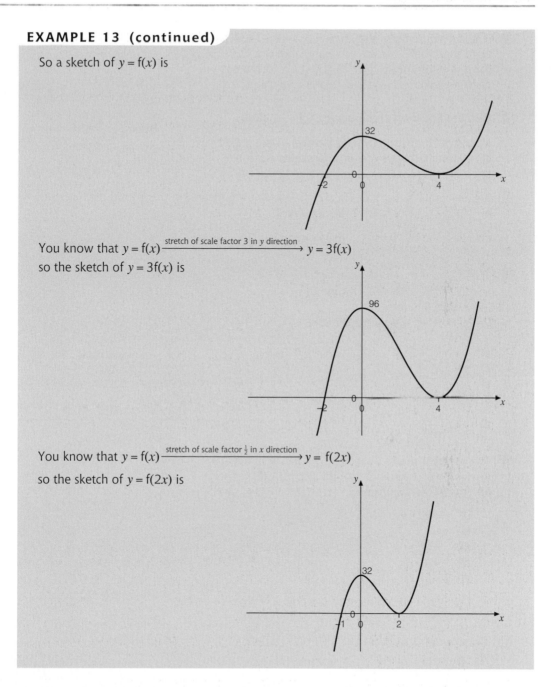

So a sketch of $y = \mathrm{f}(x)$ is

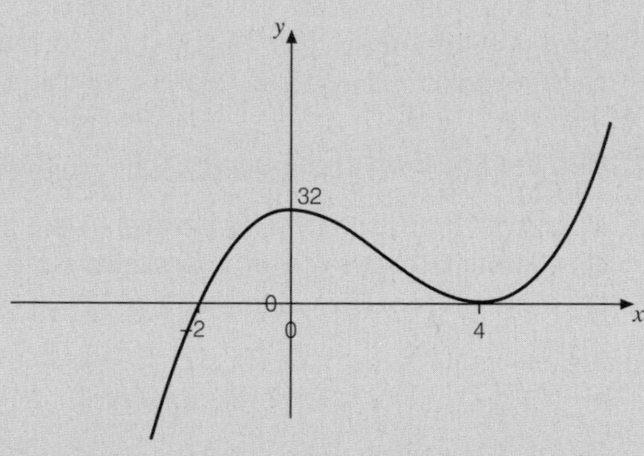

You know that $y = \mathrm{f}(x) \xrightarrow{\text{stretch of scale factor 3 in } y \text{ direction}} y = 3\mathrm{f}(x)$
so the sketch of $y = 3\mathrm{f}(x)$ is

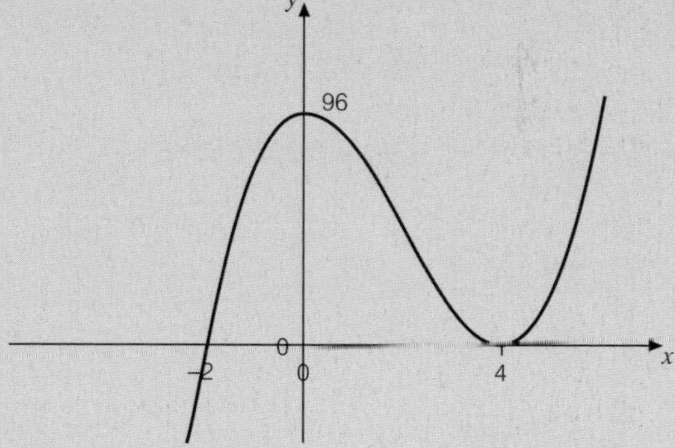

You know that $y = \mathrm{f}(x) \xrightarrow{\text{stretch of scale factor } \frac{1}{2} \text{ in } x \text{ direction}} y = \mathrm{f}(2x)$
so the sketch of $y = \mathrm{f}(2x)$ is

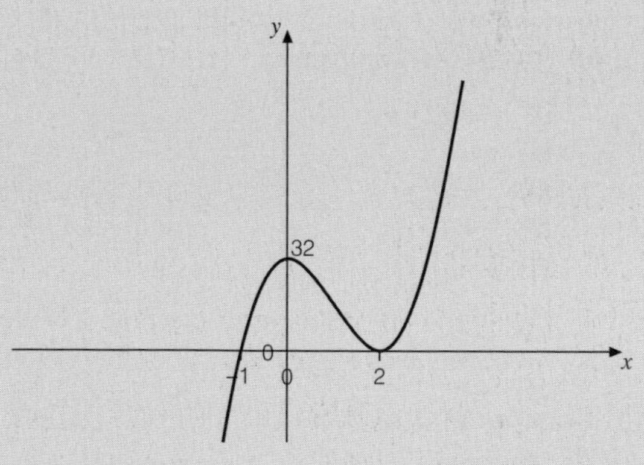

EXERCISE 5

1 Find the equation of the image of $y = x^2 + 3$ after

a) a stretch of scale factor 4 parallel to the y axis;
b) a stretch of scale factor 2 parallel to the x axis;
c) a stretch of scale factor $\frac{1}{5}$ parallel to the x axis.

2 Find the equation of the image of $y = (x - 1)(x - 4)$ after

a) a stretch of scale factor 2 parallel to the y axis;
b) a stretch of scale factor 3 parallel to the x axis;
c) a stretch of scale factor $\frac{1}{2}$ parallel to the x axis.

3 Find the equation of the image of $y = \dfrac{9}{x^2}$ after

a) a stretch of scale factor 5 parallel to the y axis;
b) a stretch of scale factor 2 parallel to the x axis;
c) a stretch of scale factor $\frac{1}{3}$ parallel to the x axis.

4 If $g(x) = (x - 4)(x - 6)(x + 2)$, sketch, on separate diagrams, the graphs of

a) $y = g(x)$ **b)** $y = 2g(x)$
c) $y = g(2x)$ **d)** $y = g\left(\dfrac{x}{3}\right)$

5 The diagram shows the graph of $y = h(x)$ for $-1 \leqslant x \leqslant 2$. Outside this interval $h(x)$ is zero.
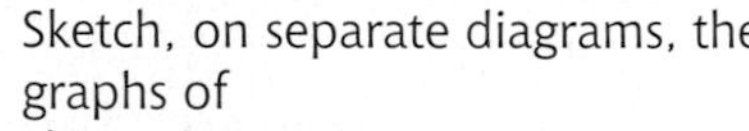
Sketch, on separate diagrams, the graphs of

a) $y = h(x) + 2$
b) $y = h(x - 3)$
c) $y = -h(x)$
d) $y = 3h(x)$
e) $y = h\left(\dfrac{x}{2}\right)$

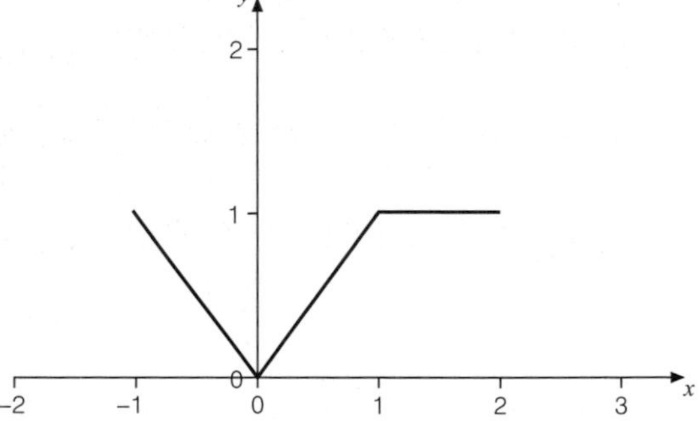

6 The diagram shows a sketch of $y = f(x)$ where $f(x) = x^2 - 6x + 8$.

a) Find the co-ordinates of the points A, B, C and D.
b) Sketch the graphs of

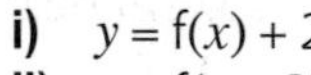

i) $y = f(x) + 2$
ii) $y = f(x + 2)$
iii) $y = 2f(x)$
iv) $y = f(2x)$
v) $y = f(-x)$

giving the co-ordinates of the points where the curves meet the axes and of the minimum points.

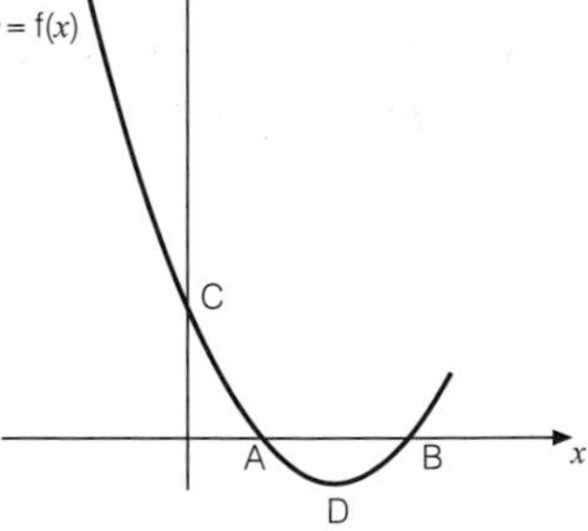

7 The diagram shows a sketch of $y = f(x)$ where

$$f(x) = x^3 - 6x^2.$$

Sketch the graphs of

i) $y = f(x) - 18$
ii) $y = f(x - 4)$
iii) $y = 3f(x)$
iv) $y = f(\frac{1}{2}x)$
v) $y = f(3x)$
vi) $y = f(-x)$

giving, where possible, the co-ordinates of the points where the curves meet the axes and of the maximum and minimum points.

Having studied this chapter you should know

- the shapes and properties of the graphs of the form $y = kx^n$
- how to sketch a graph given its rule in factorised form
- that if the factorised form contains a factor of $(x - a)^2$ then the x axis is tangential to the curve at $x = a$
- to use the six results giving the relationships between the graphs of $y = \text{f}(x)$, $y = -\text{f}(x)$, $y = \text{f}(x) + b$, $y = \text{f}(x - a)$, $y = k\text{f}(x)$, $y = \text{f}\left(\dfrac{x}{k}\right)$ and $y = \text{f}(kx)$:

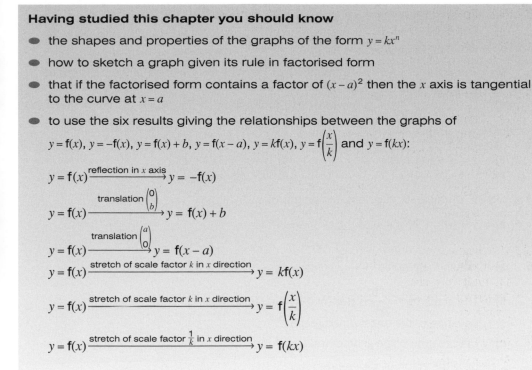

$$y = \text{f}(x) \xrightarrow{\text{reflection in } x \text{ axis}} y = -\text{f}(x)$$

$$y = \text{f}(x) \xrightarrow{\text{translation } \binom{0}{b}} y = \text{f}(x) + b$$

$$y = \text{f}(x) \xrightarrow{\text{translation } \binom{a}{0}} y = \text{f}(x - a)$$

$$y = \text{f}(x) \xrightarrow{\text{stretch of scale factor } k \text{ in } x \text{ direction}} y = k\text{f}(x)$$

$$y = \text{f}(x) \xrightarrow{\text{stretch of scale factor } k \text{ in } x \text{ direction}} y = \text{f}\left(\frac{x}{k}\right)$$

$$y = \text{f}(x) \xrightarrow{\text{stretch of scale factor } \frac{1}{k} \text{ in } x \text{ direction}} y = \text{f}(kx)$$

REVISION EXERCISE

1 **a)** Sketch the graph of $y = -2\sqrt{x}$.

b) Suggest a rule for the function whose graph is shown in the diagram below.

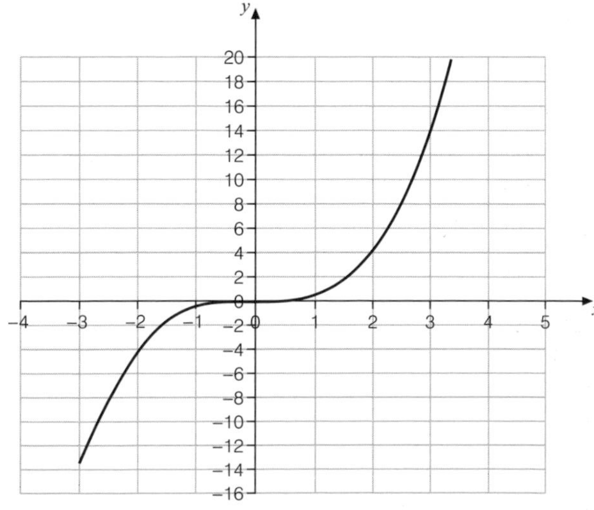

2 **a)** Sketch the graph of $y = 2\sqrt{x}$, for $x \geqslant 0$.

b) The graph of $y = 2\sqrt{x}$ is stretched by a factor of 5 parallel to the y axis. State the equation of the transformed graph.

c) The graph of $y = 2\sqrt{x}$ is translated by $\begin{pmatrix} 3 \\ 0 \end{pmatrix}$. State the equation of the transformed graph.

d) Describe the single geometrical transformation that transforms the graph of $y = 2\sqrt{x}$ onto the graph of $y = 2\sqrt{x} - 7$.

3 Sketch, on separate diagrams, the graphs of

a) $y = (x - 4)(x + 2)$ **b)** $y = (x - 4)(x + 2)^2$

c) $y = -3(x - 4)(x + 2)$ **d)** $y = 2(x - 4)^2(x + 2)$.

4 Sketch, on separate diagrams, the graphs of

a) $y = \dfrac{-6}{x}$ **b)** $y = \dfrac{120}{x^2}$ **c)** $y = \dfrac{8}{\sqrt{x}}$

5 **a)** Sketch the graph of $y = \dfrac{16}{x^2}$, for $x \neq 0$.

b) The graph of $y = \dfrac{16}{x^2}$ is stretched by a factor of $\frac{1}{2}$ parallel to the x axis. State the equation of the transformed graph.

c) Describe the single geometrical transformation that transforms the graph of $y = \dfrac{16}{x^2}$ onto the graph of $y = \dfrac{16}{(x + 5)^2}$.

6 The curve $y = k(x + a)(x + b)^2$ crosses the x axis at $(-1, 0)$, touches the x axis at $(3, 0)$, crosses the y axis at $(0, 45)$ and passes through the point $(5, p)$.

Find the values of the constants a, b, k and p.

7 If $f(x) = (3 - x)(x + 6)^2$, sketch, on separate diagrams, the graphs of

a) $y = f(x)$ **b)** $y = 2f(x)$

c) $y = f(3x)$ **d)** $y = f\left(\dfrac{x}{2}\right)$

8 The diagram shows the graph of $y = h(x)$ for $0 \leqslant x \leqslant 6$. Outside this interval $h(x)$ is zero.

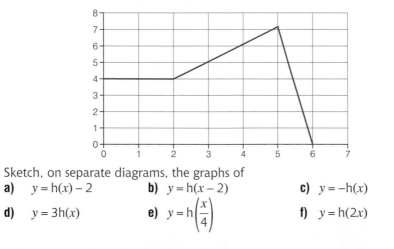

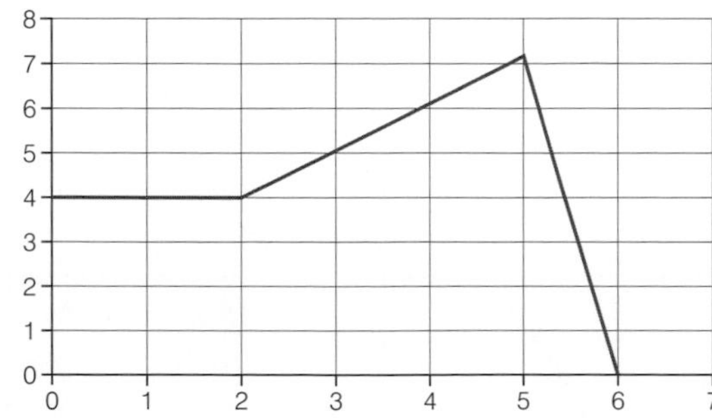

Sketch, on separate diagrams, the graphs of

a) $y = h(x) - 2$ **b)** $y = h(x - 2)$ **c)** $y = -h(x)$

d) $y = 3h(x)$ **e)** $y = h\left(\dfrac{x}{4}\right)$ **f)** $y = h(2x)$

9 Sketch the graph of $y = \frac{36}{x^2}$ for both positive and negative values of x.

Determine the gradient of $y = \frac{36}{x^2}$ at the point where $x = -2$.

Describe fully the geometrical transformation that will map the curve $y = \frac{36}{x^2}$ onto the curve $y = \frac{36}{(x+3)^2}$. Hence sketch the graph of $y = \frac{36}{(x+3)^2}$.

Use your earlier answers to deduce the gradient of $y = \frac{36}{(x+3)^2}$ at the point where $x = -5$.

REVISION 1 Indices and Surds

Revise chapters 1 and 5 before attempting this exercise. Do NOT use calculators.

1 Find the exact values of

a) 5^{-2} **b)** $16^{\frac{1}{2}}$ **c)** $(3\sqrt{2})^2$

2 Express $\dfrac{1}{(\sqrt{a})^{\frac{4}{3}}}$ in the form a^n, stating the value of n.

(OCR Jun 1995 P1)

3 Simplify

a) $(5+3\sqrt{2})(4-2\sqrt{2})$ **b)** $\dfrac{6}{\sqrt{2}}$ **c)** $\dfrac{10-6\sqrt{2}}{3-2\sqrt{2}}$

4 If $y = x^{\frac{3}{4}} + 2$

a) find the value of y when $x = 16$

b) find the value of x when $y = 29$.

5 Express $(9a^4)^{-\frac{1}{2}}$ as an algebraic fraction in simplified form.

(OCR Jun 1996 P1)

6 Find the value of x if $5^{4x+2} = 25^{x+3}$.

7 Find the exact values of

a) 7^0 **b)** $8^{\frac{2}{3}}$ **c)** $(16)^{-\frac{3}{4}}$

8 Find the values of p and q if

$$5\sqrt{p} - 3\sqrt{q} = 9$$
$$2\sqrt{p} + \sqrt{q} = 8$$

9 Evaluate

a) $8^{\frac{2}{3}} + 9^{\frac{3}{2}}$ **b)** $(\frac{9}{25})^{-\frac{3}{2}}$

10 Two straight lines have equations $y = (3\sqrt{2})x + 17$ and $y = (4\sqrt{2})x + 5$.
Find the exact co-ordinates of their point of intersection, P, giving your answer in the form $(a\sqrt{2}, b)$, where a and b are integers.

(OCR Mar 1998 P1, part)

11 If $y = (x\sqrt{x})^{-2}$ find the value of the integer n such that $y = x^n$. Hence find the co-ordinates of the point on the graph of $y = (x\sqrt{x})^{-2}$ where the gradient is $-\frac{3}{16}$.

12 Simplify $\dfrac{6+\sqrt{3}}{4-2\sqrt{3}}$ giving your answer in the form $a + b\sqrt{3}$ where a and b are rational numbers.

13 Find the exact distance between the points $(2\sqrt{5}, \sqrt{3})$ and $(-\sqrt{5}, 4\sqrt{3})$, giving your answer as a surd in its simplest form.

14 **a)** Show that $4^{x+1} \equiv 4 \times (2^x)^2$.

b) Hence solve the equation $4^{x+1} - 17 \times 2^x + 4 = 0$.

15 Find the positive root of the equation

$$x^2 - \sqrt{80}x - 55 = 0$$

expressing your answer in the form $p\sqrt{5} + q\sqrt{3}$ where p and q are integers whose values should be stated.

16 It is given that $x^n = 8$ and $x^m = 25$. Determine the numerical value of

a) x^{n+m} **b)** x^{2n} **c)** $x^{-\frac{1}{2}m}$ **d)** $\sqrt[3]{x^{n+\frac{3}{2}m}}$

REVISION 2 Polynomials

Revise chapters 2, 3, 6 and 8 before attempting this exercise. Do NOT use calculators.

1 Expand and simplify $(x + 1)(x^2 - 2x + 2)$.

(OCR Mar 1999 P1)

2 Solve the simultaneous equations

$2x + y = 3 \qquad 2x^2 - xy = 10$

(OCR Jun 1995 P1)

3 Solve the inequalities

a) $24 - 5x < 9$ **b)** $-7 \leqslant 3x + 2 < 20$

4 Find the exact solutions of the equation $x^2 - 12x - 5 = 0$, expressing your answers as simply as possible.

5 Factorise $(x + 3)^2 - 16$.

(OCR Nov 1996 P1)

6 The quadratic equation $x^2 + kx + 36 = 0$ has two different real roots. Find the set of possible values of k.

(OCR Mar 1996 P1)

7 A biologist claims that the average height, h metres, of trees of a certain species after t months' growth is given by

$h = \frac{1}{5}t^{\frac{2}{3}} + \frac{1}{8}t^{\frac{1}{3}}$.

i) Find the average height of trees of this species after 64 months.

ii) Find the number of months that the trees have been growing when the average height is $\frac{21}{20}$ metres.

(OCR Mar 1996 P1, adapted)

8 Solve the simultaneous equations

$x + y = 1 \qquad x^2 - xy + y^2 = 7$

(OCR Jan 1996 P1)

9 a) Solve the equation $x^2 - 8\sqrt{5}x + 35 = 0$, giving your answers in terms of surds, simplified as far as possible.

b) Hence solve the inequality $x^2 - 8\sqrt{5}x + 35 \geqslant 0$.

10 Find the set of values of a for which the equation $ax^2 - 6x + a = 0$ has two distinct real roots.

(OCR Nov 1996 P1)

11 Solve the inequality $x^2 - 13x + 12 < 0$.

(OCR Feb 1997 P1)

12 i) Write $x^2 + 6x - 13$ in the form $(x + p)^2 + q$ stating the values of the constants p and q.

ii) Hence, or otherwise, find the exact solutions of the equation $x^2 + 6x - 13 = 0$.

iii) Solve the inequality $x^2 > 13 - 6x$.

13 If $f(x) = x^3 + 2x^2 - 4$ and $g(x) = x^2 + 5$, simplify

a) $f(x) + 2g(x)$ **b)** $f(x)g(x)$ **c)** $g(2x - 5)$

14 Solve the simultaneous equations

$$2x - y = 1, \qquad 4x^2 + y^2 = 13$$

(OCR Nov 1997 P1)

15 Factorise $49x^2 - 21x + 2$.
Hence, or otherwise, solve the equation $49y - 21\sqrt{y} + 2 = 0$.
Give your answers as fractions.

(OCR Jun 1998 P1)

16 **a)** **i)** Find the discriminant of the quadratic polynomial $5x^2 + 8x + 9$.

ii) How many real roots does the equation $5x^2 + 8x + 9 = 0$ have?

iii) What can now be deduced about the graph of $y = 5x^2 + 8x + 9$?

b) **i)** Find the value of k if the equation $5x^2 + 8x + k = 0$ has one real root.

ii) Sketch the graph of $y = 5x^2 + 8x + k$ for this value of k. State the co-ordinates of the vertex of the graph.

17 Find the values of the constants p, q and r if

$$(2x - 1)^2 + (x + 3)^2 \equiv px^2 + qx + r.$$

18 **i)** Express $2x^2 + 8x - 3$ in the form $a((x + p)^2 + q)$, stating the values of the constants a, p and q.

ii) Sketch the graph of $y = 2x^2 + 8x - 3$, stating the co-ordinates of the vertex.

19 Factorise completely

a) $6x^3 - 54x$ **b)** $10x^2 - 26x - 12$

20 Solve the inequality $3x^2 + x - 2 > 0$.

(OCR Nov 1997 P1)

21 Solve the simultaneous equations

$$x^2 + 2y^2 = 9 \qquad x + 4y = 9$$

(OCR Jun 1998 P1)

22 The diagram shows the graph of $y = x^2 - 2px + p$, where p is a positive constant.
The point A is the lowest point on the graph and is given to lie above the x axis.

i) By completing the square, express the co-ordinates of A in terms of p.

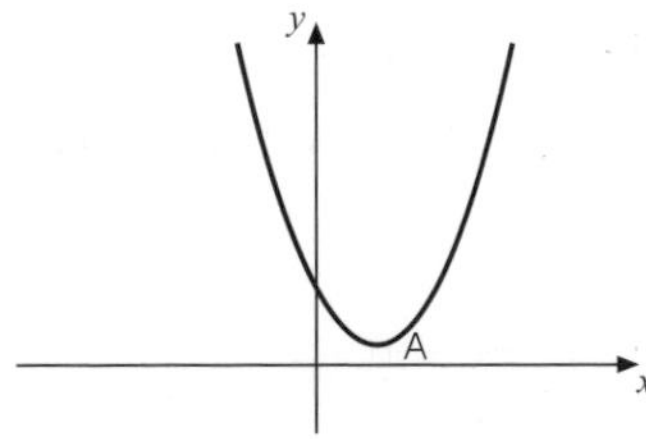

ii) Show that p must satisfy the inequality $p - p^2 > 0$ and hence find the set of possible values of p.

iii) Given that A also lies on the straight line with equation $y = 2x - 1$, find the exact value of p.

(OCR Jun 1998 P1)

23 **i)** Given that $u^{\frac{1}{3}} = z$ show that the equation $2u^{\frac{1}{3}} + 3u^{-\frac{1}{3}} = 7$ can be written as $2z^3 - 7z + 3 = 0$.

ii) Hence solve the equation $2u^{\frac{1}{3}} + 3u^{-\frac{1}{3}} = 7$.

24 Express $-3x^2 + 12x + 5$ in the form $a(x + p)^2 + q$ stating the values of the constants a, p and q.
Hence sketch the graph of $y = -3x^2 + 12x + 5$, stating the co-ordinates of the vertex.

REVISION 3 Co-ordinate Geometry and Graphs

Revise chapters 4, 10 and 11 before attempting this exercise. Do NOT use calculators.

1 The co-ordinates of A and B are (2, 3) and (4, −3), respectively. Find the length of AB and the co-ordinates of the mid-point of AB.

(OCR Nov 1995 P1)

2 The points A and B have co-ordinates (1, 5) and (3, 1), respectively. Find the equation of the perpendicular bisector of AB.

(OCR Nov 1996 P1)

3

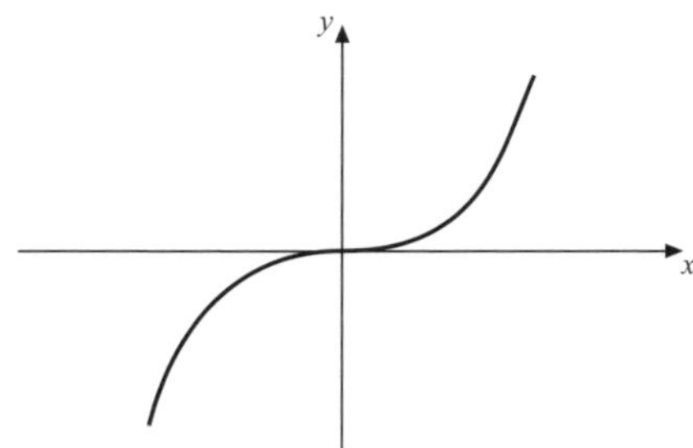

The diagram shows a sketch of the graph of $y = kx^n$, where k and n are integers. The graph passes through the point (2, 160).
Write down possible values for the constants k and a.

4 Find the equation of the circle which has the points (3, 7) and (−5, 13) at opposite ends of a diameter.
The tangent to this circle at (3, 7) meets the x axis at P and the y axis at Q. Prove that the triangle OPQ has area $\frac{27}{8}$ units^2.

5 Find the equation of the line, L_1, passing through the points (7, 11) and (11, 5), giving your answer in the form $ax + by + c = 0$ where a, b and c are integers.
Find the equation of the line L_2 which is perpendicular to L_1 and passes through the point (1, 6), giving your answer in the form $px + qy + r = 0$ where p, q and r are integers.

6 Without using differentiation, sketch the graph of $y = (2x - 5)(x + 4)(x - 6)$ and hence solve the inequality $0 \leqslant (2x - 5)(x + 4)(x - 6)$.

7 Solve the simultaneous equations

$$y = 2x^2 + x - 5$$
$$y = 9x - 13$$

and interpret your solution geometrically.

8 The diagram, not drawn to scale, shows a trapezium OABC with OA parallel to CB. Given that B is the point (4, 3), C is the point (0, 2) and the diagonal CA is parallel to the x axis, **calculate** the co-ordinates of A.

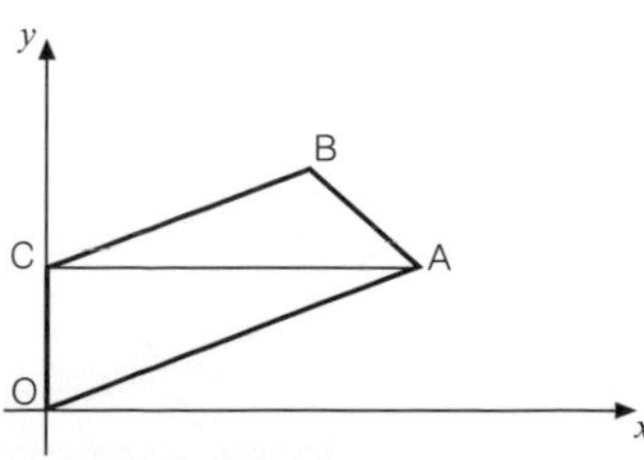

(OCR Jun 1996 P1)

9

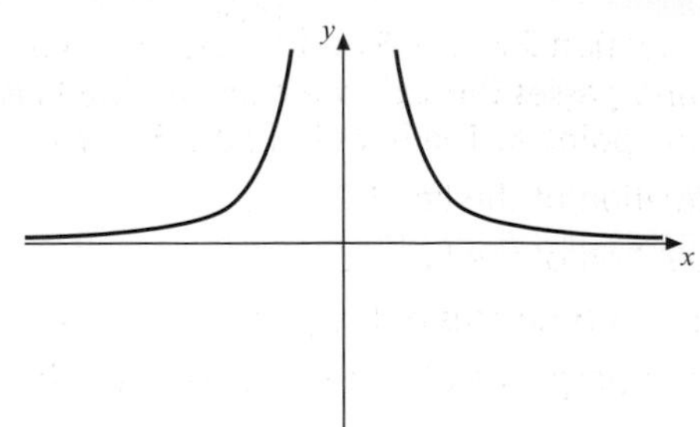

The function h is given by $\mathrm{h}(x) = \dfrac{k}{x^n}$ where k and n are positive integers.

The graph of $y = \mathrm{h}(x)$ is shown in the diagram. The points (1, 100) and (2, 25) lie on the graph.

i) Determine the values of k and n.

ii) Sketch the graphs of
a) $y = -\mathrm{h}(x)$ **b)** $y = \mathrm{h}(x - 2)$ **c)** $y = \mathrm{h}(x) + 3$

10 The straight line L has equation $2x - 5y = 7$ and the point A has co-ordinates (3, −2).

i) Find the gradient of L.

ii) Find the equation of the straight line that passes through A and is perpendicular to L. Give your answer in the form $ax + by + c = 0$, where a, b and c are integers.

(OCR Mar 1999 P1)

11 **a)** Find the centre and radius of the circle whose equation is
$$x^2 + y^2 - 10x + 4y + 4 = 0.$$

A second circle has centre (14, 10) and touches the first circle externally at a point C.

b) Find the equation of the second circle.

c) Determine the co-ordinates of the point C.

d) Find the equation of the line through C which is tangential to both circles.

12 Find the centre and radius of the circle $x^2 + y^2 + 8x - 6y - 144 = 0$.
Show that the point A(7, 12) lies outside this circle.
The point P moves around the circle $x^2 + y^2 + 8x - 6y - 144 = 0$. Determine the greatest possible distance of the point P from the point A.
The point T lies on the circle $x^2 + y^2 + 8x - 6y - 144 = 0$ and is such that the line AT is a tangent to the circle. Calculate the length of AT.

13

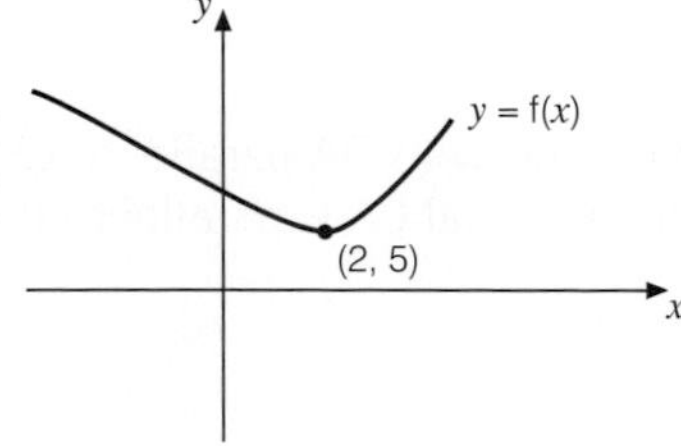

The only stationary point on the curve $y = \mathrm{f}(x)$ has co-ordinates (2, 5), as shown in the diagram. State the co-ordinates of the stationary point of each of the following curves:

i) $y = \mathrm{f}(x + 1)$ **ii)** $y = \mathrm{f}(-x)$

iii) $y = 3\mathrm{f}(x)$ **iv)** $y = \mathrm{f}(4x)$

(OCR Jun 2000 P2, adapted)

14 The line L_1 has equation $2y - x = 6$ and the point A has co-ordinates $(1, -4)$. The line L_2 is parallel to L_1 and passes through the point A. The line L_3 is perpendicular to L_1 and passes through the point A. The lines L_3 and L_1 intersect at B.

i) Find the equation of the line L_2.

ii) Find the equation of the line L_3.

iii) Calculate the co-ordinates of the point B.

iv) Prove that the perpendicular distance between the lines L_1 and L_2 is $3\sqrt{5}$ units.

15 The diagram shows the graph of $y = g(x)$.

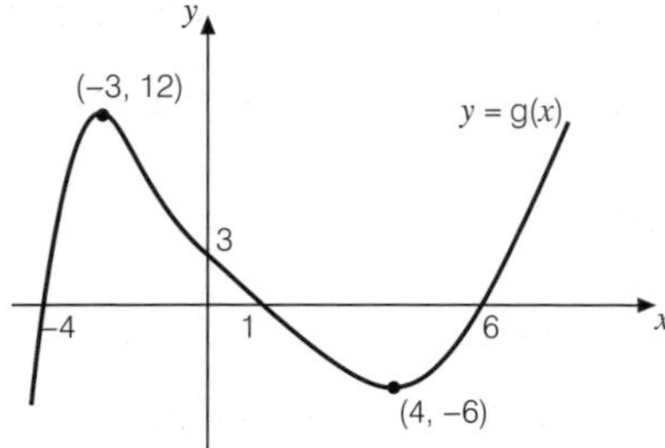

Sketch the graphs of

a) $y = -g(x)$ **b)** $y = 2g(x)$ **c)** $y = f\left(\frac{x}{2}\right)$

taking care to show, in each case, the co-ordinates of the stationary points and the intercepts with the co-ordinate axes.

16 Find the intersection of the graphs $y = -2x + 5$ and $(x + 2)^2 + (y + 1)^2 = 20$ and interpret your answer geometrically.

17 The points A, B, C and D form a rhombus. The points B and D have co-ordinates (4, 0) and (0, 2), respectively.

a) Find the co-ordinates of the mid-point and the gradient of the line segment joining B and D.

b) Prove that the equation of AC is $y = 2x - 3$.

The side CD of the rhombus has equation $y = x + 2$.

c) Calculate the co-ordinates of the points C and A.

d) Find the length of the perimeter of the rhombus, giving your answer in surd form, as simply as possible.

18 A function f is given by $f(x) = k(x + a)(x + b)^2$, where k, a and b are constants. The diagram shows the graph of $y = f(x)$.

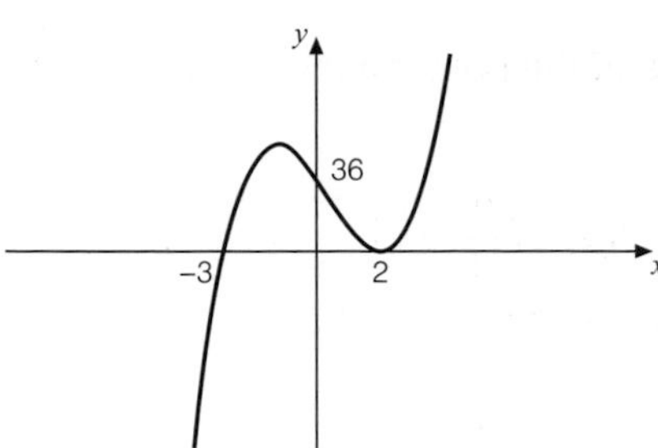

i) Find the values of the constants a, b and k.

ii) Sketch the graphs of

a) $y = f(x + 2)$ **b)** $y = f\left(\frac{x}{2}\right)$

19 The points P and Q are (2, 8) and (6, 0), respectively.
Find the equation of the perpendicular bisector of PQ.
Deduce the equation of the circle that has its centre on the y axis and passes through the points P and Q.

20 Sketch the graph of $y = 3\sqrt{x}$.
Calculate the co-ordinates of the points of intersection of the curve $y = 3\sqrt{x}$ and the line $y = x + 2$.

21 Two straight lines have equations $x + 3y = 13a$ and $3x - y = 9a + 10$, where a is a constant. Their point of intersection is denoted by A.

i) Find, in terms of a, the co-ordinates of A.

ii) Show that, for all values of a, the point A lies on the straight line with equation $3x - 4y = 13$.

(OCR Mar 1999 P1)

22

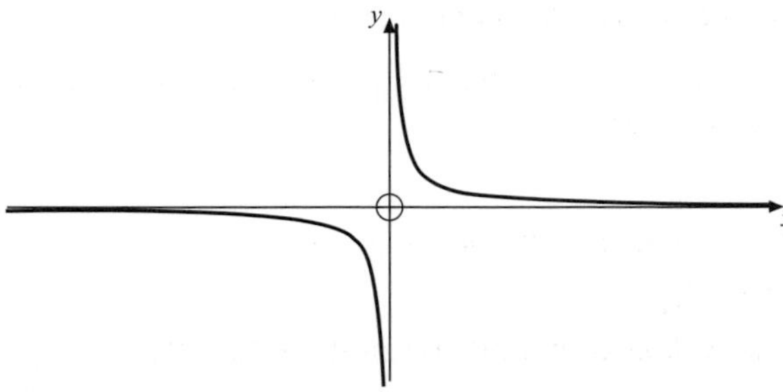

The diagram shows a sketch of the graph of $y = kx^n$, which passes through the point (1, 3). Find the value of k.
By considering the form of the graph, write down three **possible** values for n.
Given further that the point $(2, \frac{3}{8})$ also lies on this graph, find the value of n.

(OCR Nov 1996 P1)

REVISION 4 Differentiation

Revise chapters 7 and 9 before attempting this exercise. Do NOT use calculators.

1 Find the x co-ordinate of the point on the curve $y = x^{\frac{3}{2}}$ at which the gradient is 9.

(OCR Jun 2000 P2)

2 If $z = \sqrt{t} + \frac{4}{t}$ find the values of $\frac{dz}{dt}$ and $\frac{d^2z}{dt^2}$ when $t = 4$.
Explain the significance of your answers.

3 **a)** Find $\frac{dy}{dx}$ if $y = (x^2 + 1)(3x - 2)$.

b) Find $g'(t)$ if $g(t) = \frac{t^5 + 6}{t}$.

4 Find by differentiation the exact co-ordinates of the stationary point of the curve with equation $y = \frac{1}{x} + \frac{6}{x^2}$.
Determine whether this stationary point is a maximum or a minimum, showing your working.

(OCR Jun 2000 P2)

5 Find the equation of the normal to the curve $y = 2x^{\frac{1}{3}}$ at the point on the curve where $x = 8$. Give your answer in the form $y = mx + c$.

(OCR Nov 1999 P2)

6 The area of a puddle t hours after the rain storm finished is given by

$$A = \tfrac{1}{100}(t + 10)(8 - t) \qquad 0 < t < 8.$$

Find the value of $\frac{dA}{dt}$ when $t = 3$ and state its significance.

7 Find the exact co-ordinates of the stationary point on the curve $y = x^{\frac{3}{2}} - 6x$. Show that this stationary point is a minimum point.

8 Find the equation of the tangent to the curve $y = 2x^2 - 3x + 9$ at the point $(-1, 14)$. Give your answer in the form $y = mx + c$.

(OCR Mar 1999 P2)

9 A function f is defined by $f(x) = x^3 - 9x^2 - 120x + 30$.

a) Evaluate $f'(3)$.

b) For what values of x is the function decreasing?

10 Find by differentiation the co-ordinates of the stationary point on the curve with equation

$$y = 2x^2 + 8x + 13.$$

(OCR Jun 1997 P2)

11 It is given that $y = k\sqrt{x}$, where k is a constant and $x > 0$. When $x = 3$, the value of $\frac{dy}{dx}$ is $\sqrt{3}$. Find the value of k.

(OCR Nov 1995 P2)

12 The diagram shows an open rectangular tank, of height h meters, with a horizontal square base of side x metres. The tank can hold a volume of 13.5 m^3 of water and the internal surface area of the tank is S m^2.

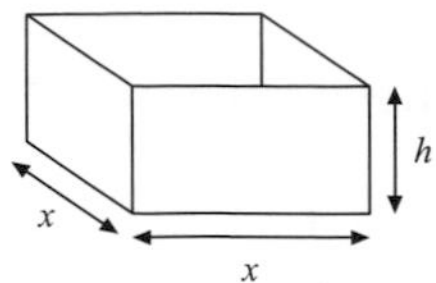

a) Show that $S = x^2 + \dfrac{54}{x}$.

b) Differentiate S with respect to x and hence find the dimensions of the tank when S is a minimum. Show clearly that, in this case, S is a minimum and not a maximum.

(OCR Mar 98 P2)

13 Find the values of x for which the function $f(x) = x^{\frac{4}{3}} - 2x + 5$ is increasing.

14 A curve has equation

$$y = 3px^2 - x^3, \qquad x > 0,$$

where $p > 0$.
Use differentiation to prove that the co-ordinates of the stationary point are $(2p, 4p^3)$.
Determine whether the stationary point is a maximum or a minimum point.
Hence write down the maximum value of $y = 96x^2 - x^3$ for positive values of x, giving your answer as a power of 2.

15 A new make of personal organiser costs £100 each to manufacture and market. Market research indicates that the number, N thousand, of organisers that will be sold in the first year is given by $N = aS + b$, where a and b are constants and S is the selling price. The numbers that will be sold at two particular selling prices are as given in the table.

S	N
120	30
140	10

a) Find the values of a and b.

b) Express the profit per organiser in terms of S and hence show that the total profit in the first year is £T thousand where

$$T = -15\,000 + 250S - S^2$$

c) Find the selling price which will produce the maximum possible total profit in the first year and find this total profit. (You should show that this profit is a maximum and not a minimum.)

(OCR Mar 1996 P2)

16 The force, F Newtons, of attraction between two particles that are a distance x metres apart is given by

$$F = \frac{0.2}{x^2}.$$

Find the rate of change of F with respect to x when $x = 20$, giving your final answer in standard form.

17 Find the equation of the normal to the curve $y = \dfrac{16}{x}$ at the point P(2, 8). The normal intersects the curve again at the point Q. Find the co-ordinates of Q.

SAMPLE PAPER

1 Expand the brackets and simplify

a) $(x+2)^2-(x-1)^2$ [3]

b) $(2x+1)(x^2-3x+4)$ [2]

2 **a)** Evaluate $16^{-\frac{3}{4}}$ [2]

b) Simplify $4x^{\frac{5}{2}} \times 5x^{-2}$ [2]

c) Express $\dfrac{4-\sqrt{2}}{2+\sqrt{2}}$ in the form $a+b\sqrt{2}$ where a and b are integers. [3]

3 **i)** Sketch the curve $y=\dfrac{1}{x}$. [1]

ii) The curve $y=\dfrac{1}{x}$ is translated by 4 units parallel to the x-axis.
Find the equation of the curve after this transformation. [2]

iii) Describe fully the transformation that transforms the curve $y=\dfrac{1}{x}$ to the curve $y=\dfrac{3x+1}{x}$. [3]

4 Given that $f(x)=2x^3+\dfrac{1}{\sqrt[3]{x}}-1$, find

i) $f'(x)$, [3]

ii) $f''(x)$ [2]

5 Find the range of values of x for which the inequalities
$2x+6>1-3x$ and $x^2 \geqslant x$ are both valid. [6]

6 **i)** Given that $3x^2-6x+2=a(x+b)^2+c$ for all values of x, find the constants a, b and c. [4]

ii) Hence, or otherwise, solve the equation $3x^2-6x+2=0$, giving your answers in the form $p+q\sqrt{3}$, where p and q are rational numbers. [4]

7 **i)** Find the gradient of the line L whose equation is $x-2y=7$ [1]

ii) Find the equation of the line through the point (1, 3) which is parallel to the line L. Give your answer in the form $px+qy=r$, where p, q and r are integers. [3]

iii) The points A(a, −1) and B(9, b) lie on L. Show that the perpendicular bisector of the line joining A and B passes through (3, 8). [6]

8 **i)** Find the coordinates of the stationary points on the curve $y=x^3-6x^2+9x$. [6]

ii) Determine whether each stationary point is a maximum or a minimum point. [3]

iii) Sketch the curve $y=x^3-6x^2+9x$, showing clearly the points where it meets the axes. [3]

9 **i)** Find the values of k for which the line $y = x + k$ is a tangent to the circle $x^2 + y^2 = 9$. [5]

ii) Find the perpendicular distance between the two tangents. [2]

iii) Prove that the point (1, 2) lies inside the circle $x^2 + y^2 = 9$ but that it lies outside the circle

$$x^2 + y^2 - 4x + 6y - 3 = 0.$$ [6]

ANSWERS

CHAPTER 1
Exercise 1

1 i) 16 **ii)** 243 **iii)** 64
iv) −8 **v)** 81 **vi)** 100 000

2 i) 19.45 **ii)** 2877.94 **iii)** 0.42
iv) −21.67 **v)** 5479.70

3 a) 2^{11} **b)** 3^6 **c)** 5^{21}
d) 2^{13} **e)** 7^9 **f)** 5^7
g) 4^8 **h)** 5^{20} **i)** 3^9

4 a) a^{16} **b)** y^4 **c)** c^{18} **d)** z^8
e) y^4 **f)** y^{24} **g)** t^{25} **h)** t^{15}
i) p^{20} **j)** x^5 **k)** d^2 **l)** t^2

5 a) $20s^5t^8$ **b)** $3s^6t$ **c)** $4st^2$
d) $24s^3t^6u^5$ **e)** $30s^6t^7$ **f)** $2s^3t$

CHAPTER 1
Exercise 2

1 a) $\frac{1}{8}$ **b)** 3 **c)** $\frac{1}{5}$ **d)** 216
e) $\frac{1}{2}$ **f)** 4 **g)** 25 **h)** $\frac{1}{4}$
i) $\frac{1}{343}$ **j)** $\frac{1}{8}$ **k)** $\frac{25}{9}$ **l)** $\frac{8}{27}$
m) $\frac{9}{4}$ **n)** $\frac{8}{27}$ **o)** $\frac{8}{125}$

2 a) $81x^8$ **b)** $\dfrac{8x^3}{y^6}$ **c)** $64p^6q^3$
d) $\dfrac{4p^2}{q^3}$ **e)** $\dfrac{1}{16y^2}$ **f)** $\dfrac{64x^6}{1296x^4} = \dfrac{4x^2}{81}$
g) $5x^2$ **h)** $\dfrac{1}{y^{18}}$ **i)** $9y^4$
j) $\dfrac{1}{2p^2}$

3 a) x **b)** x^2 **c)** $8x^{\frac{1}{2}}$ **d)** p^3
e) $8y^4$ **f)** x^3 **g)** $x^{\frac{1}{2}}$ **h)** 1

4 a) x **b)** $8x^2$ **c)** y^2

CHAPTER 1
Exercise 3

1 a) $p = 4$ **b)** $y = 2$ **c)** $t = -3$
d) $t = 5$ **e)** $x = \frac{1}{2}$ **f)** $p = 2$
g) $t = 1$ **h)** $p = 2$ **i)** $s = 3$
j) $t = 3$

2 a) 4.16 **b)** ±2.24 **c)** 4.35
d) 0.936 **e)** ±0.900

3 a) $u = 64$ **b)** $v = \frac{1}{81}$ **c)** $s = -32$
d) $t = 32$ **e)** $z = \frac{1}{2}$ **f)** $y = \frac{1}{81}$

CHAPTER 1
Revision Exercise

1 a) p^8 **b)** p^{28} **c)** p^{-7} or $\dfrac{1}{p^7}$

2 a) $\frac{1}{4}$ **b)** 1 **c)** 8

3 a) $\dfrac{y^{10}}{2}$ **b)** $3y^4$ **c)** $\dfrac{1}{25y^{12}}$

4 a) x **b)** 1

5 a) $t = -2$ **b)** $t = 2$ **c)** $y = 1$

6 a) ±1.43 **b)** $r = 1.88$ **c)** $s = 0.758$

7 i) $\frac{16}{9}$ **ii)** $2x$

8 a) 64 **b)** $u = \frac{1}{10}$ **c)** $v = \pm 32$

9 a) 6 **b)** $\frac{25}{9}$ **c)** $\frac{1}{8}$ **d)** −32

10 a) x^5 **b)** x^3 **c)** x **d)** $8x^5$

11 i) 45 **ii)** $\frac{2}{5}$ **iii)** $\frac{25}{3}$

12 a) 4 **b)** $t = \pm 9$

CHAPTER 2
Exercise 1

1 i) a) 10 **b)** 5 **c)** 1.09 **d)** 145

2 i) a) 64 **b)** −8
c) 1000 **d)** 0.216
ii) $8x^3$ $(x+2)^3$
ii) $\alpha = 5$ $\beta = -0.2$

3 i) a) 6 **b)** 48 **c)** $\frac{5}{4}$
ii) a) $48x^2 + 20x - 2$
b) $12x^2 - 2x - 4$

CHAPTER 2
Exercise 2

1 a) deg f = 4 deg g = 3
deg h = 4 deg j = 3
b) i) 3 **ii)** 3 **iii)** −2 **iv)** 0

$x - 7$
$- 2x - 7$
$6x^4 + 14x^3$
$- 5x^3 + 13x - 13$

$5x^3 + 2x - 22$
g) $2x^7 + x^5 + 6x^4 - 14x^3 + 3x^2 - 7x$
h) $6x^4 - 4x^3 + 17x + 2$
i) $6x^7 - 10x^6 + 3x^5 + 9x^4 + 2x^3 + 7x^2 + x$

3 **a)** 2 **b)** 1

4 **i)** deg pq = deg p + deg q
ii) deg (p + q) ⩽ max(deg p, deg q)

CHAPTER 2
Exercise 3

1 **i)** $6p^2 + 21p$ **ii)** $15p^2 - 35p$
iii) $5x^5 + 30x^2$ **iv)** $8y^2 + y$
v) $15x^2 + 2x - 8$
vi) $2p^3 + 4p^2 - 4p - 8$
vii) $p^2 + 5p - 36$
viii) $y^2 - 17y + 72$
ix) $10t^2 + 39t + 35$ **x)** $4t^2 - 81$
xi) $9t^2 - 6t + 1$ **xii)** $2x^2 - 114$
xiii) $2x^2 + 18$
xiv) $3x^3 + 8x^2 - x - 10$

2 **a)** 12
b) $5x^8 + 3x^4 - 2$ degree = 8
d) $a = 2$ $b = -5$

4 **a)** $2x^4 - 7x^3 - 15x^2$ **b)** $14T$
c) $29T$

CHAPTER 2
Exercise 4

1 $x(x + 9)$ **2** $3x(x - 2)$
3 $p(p - 5)$ **4** $6p(p - 3)$
5 $x^3(2x^2 + 7)$ **6** $6(x^4 + 2x^2 + 4)$
7 $3x^3(3x^2 - 2x + 7)$ **8** $4y^5(2 - 3y)$
9 $4s^2(1 + 2s^2)$ **10** $6t^2(2t - 3)$

CHAPTER 2
Exercise 5

1 $(x + 4)(x + 7)$ **2** $(x + 4)(x + 8)$
3 $(x + 3)(x + 4)$ **4** $(x + 5)(x + 15)$
5 $(x + 7)(x - 4)$ **6** $(x - 5)(x + 3)$
7 $(x + 10)(x - 2)$ **8** $(x - 10)(x + 3)$
9 $(x - 5)(x - 6)$ **10** $(x - 2)(x - 6)$
11 $(x - 2)(x - 7)$ **12** $(x + 15)(x - 4)$
13 $(x + 4)^2$ **14** $(x + 8)(x - 8)$
15 $(x + 6)(x - 5)$ **16** $(x + 7)(x - 6)$
17 $(x - 8)(x + 6)$ **18** $(x + 10)(x - 7)$
19 $(x + 5)(x - 5)$ **20** $(x + 5)(x + 7)$
21 $(x + 10)^2$ **22** $(x - 6)^2$
23 $(x + 13)(x - 3)$ **24** $(x - 9)(x + 9)$
25 $(x - 10)(x - 8)$ **26** $(x - 11)(x + 11)$
27 $(x + 14)(x - 4)$

CHAPTER 2
Exercise 6

1 $3x(x + 6)(x - 1)$ **2** $4x(x + 3)(x - 3)$
3 $p^2(p - 2)(p - 5)$ **4** $5p(p + 6)(p - 6)$
5 $4y^3(y + 2)(y - 2)$ **6** $6x(x - 2)(x - 10)$
7 $8(p - 5)(p + 3)$ **8** $7z^2(z - 2)(z + 2)$
9 $2y(y + 4)(y + 5)$ **10** $5t(t + 8)(t - 5)$
11 $5t^5(t + 7)(t - 1)$ **12** $9x^2(x + 3)(x - 3)$

CHAPTER 2
Exercise 7

1 $(2x + 1)(x + 2)$ **2** $(2x + 1)(x + 3)$
3 $(3t + 2)(t + 1)$ **4** $(2y + 3)(y - 5)$
5 $2(x + 1)(x + 3)$ **6** $(5x + 4)(x - 1)$
7 $(7x - 3)(x - 4)$ **8** $3(p - 7)(p + 2)$
9 $(5x + 2)(x - 3)$ **10** $(11x + 3)(x - 2)$
11 $(3y - 1)(2y - 3)$ **12** $(4x - 1)(x + 2)$
13 $(3v - 1)(v - 2)$ **14** $(5s - 3)(2s + 5)$
15 $(3w - 7)(2w + 5)$ **16** $6(x - 1)(x - 2)$
17 $2(2c - 5)(c + 3)$ **18** $3(2u + 1)(u + 1)$
19 $(7x - 1)(3x - 2)$ **20** $(3x - 5)(4x - 3)$
21 $8x(x - 2)$ **22** $3x(2x + 3)$

23 $(5x+3)(x-3)$ **24** $(3a+4)(2a+3)$

25 $4(x-2)(x-4)$ **26** $2x(8x-9)$

27 $2(x-1)(4x-5)$

CHAPTER 2

Exercise 8

1 $x(x+2)(x+5)$

2 $(x-2)(x+2)(x-3)(x+3)$

3 $x(x-1)(2x+11)$

4 $-(y-3)(y+3)(y^2+3)$

5 $(x-1)(x+1)(3x^2-2)$

6 $-5x^3(x-7)(2x+3)$

7 $-(x+8)(x-2)$

8 $-(x+3)(x-3)(x^2+4)$

9 **a)** $(x^4+4)(x^4+5)$
b) $(3x^4+5)(x^4+2)$
c) $(x-2)(x+2)(x^2+4)(x^4+2)$
d) $(x-1)(x+1)(x^2+1)(x^4+1)$

10 **a)** $(x^3+1)(x^3+8)$
b) $(2x^5-1)(x^5+3)$
c) $3x^4(2x^4+7)(x^4+1)$

11 **a)** $(x^3-12)(x^3+8)$ **b)** 260 000

12 **a)** $(y^4-6)(y^4+4)$ **b)** 65 000

CHAPTER 2

Revision Exercise

1 $x(x-4)(x+4)$

2 x^3-x^2+2

3 **a)** 11 **b)** $4x^3-2x^2+2$
c) $10x^2-22$
d) $3x^6+2x^5-8x^4-8x^3+34x^2-35$

4 $p(7p-13)$

5 **a)** $6p^2+11p-10$ **b)** $9p^2-12p+4$
c) $9u^2-25$ **d)** $2x^2+24x+72$
e) $20x$

6 **a)** $3p(2p-3)$ **b)** $(p+9)(p-2)$
c) $(y+4)(y-4)$ **d)** $(y-5)(y-6)$
e) $5(t+2)(t-7)$
f) $(x^2+9)(x^2-4)=(x^2+9)(x+2)(x-2)$

7 **a)** $(3p+13)(p-2)$ **b)** $(5p+1)(p+2)$
c) $5y(y+3)(y-1)$

8 $2y^3+y^2-5y+2$

9 **a)** $(x+3)(x+8)$ **b)** 995 000

10 **a)** $12x^3-18$
b) $10x^4-4x^3-15x+6$
c) $4x^6-12x^3+9$

11 **a)** $6p(p-2)(p+2)$
b) $9p^2(2-3p)(2+3p)$
c) $(2y^2+3)(y+2)(y-2)$
d) $(t^3+12)(t^3-2)$
e) $-(x^2+16)(x+2)(x-2)$

13 $a=2$ $b=2$ $c=-18$

14 $p=3$ $q=-4$ $r=14$

CHAPTER 3

Exercise 1

1 $x=\pm 8$ **2** $p=\pm 4$ **3** $y=\pm 7$

4 $t=\pm 9$ **5** $u=\pm 3$ **6** $p=\pm 6$

CHAPTER 3

Exercise 2

1 **a)** $x=6$ or -1 **b)** $x=-\frac{1}{2}$ or 2
c) $x=0$ or 9 **d)** $x=4$ or -3
e) $t=\frac{1}{3}$ or $\frac{1}{2}$ **f)** $p=\frac{2}{3}$ or -1

2 **a)** $x=6$ or 10 **b)** $u=-\frac{15}{4}$ or 2

3 **a)** $x=20$ or -4 **b)** $x=0$ or 7
c) $x=-11$ or 7

4 **a)** $3(x+7)(x-2)$ **b)** $x=-7$ or 2

5 **a)** $5(x-3)(x+3)$ **b)** $x=\pm 3$

6 **a)** $x=-4$ or 2 **b)** $x=9$ or -3
c) $x=-9$ or 2 **d)** $y=1$ or $-\frac{13}{5}$
e) $q=\frac{3}{2}$ or 2 **f)** $x=\frac{1}{2}$ or -3

7 **a)** $x=-\frac{5}{2}$, 5 **b)** $r=\frac{7}{4}$ or 1
c) $x=0$ or $\frac{22}{3}$ **d)** $x=-\frac{1}{7}$ or 3
e) $t=4$ or -1

8 **a)** $a=2$ $-\frac{5}{2}$ **b)** $c=-35$ $-\frac{7}{2}$

CHAPTER 3
Exercise 3

1 a) $(x+5)^2-25$ **b)** $(x+2)^2-4$
c) $(x+7)^2-49$ **d)** $(x+\frac{5}{2})^2-\frac{25}{4}$
e) $(x+\frac{7}{2})^2-\frac{49}{4}$ **f)** $(x-5)^2-25$
g) $(x-8)^2-64$ **h)** $(x-\frac{9}{2})^2-\frac{81}{4}$
i $(x-10)^2-100$ **j)** $(x+6)^2-36$
$x^2+px=(x+\frac{1}{2}p)^2-\frac{1}{4}p^2$

2 a $x=-4\pm\sqrt{31}$ **b)** $x=-9\pm\sqrt{116}$
c) $x=6\pm\sqrt{27}$ **d)** $x=\frac{7}{2}\pm\sqrt{\frac{41}{4}}$

3 $(x-4)^2-11$ $x=4\pm\sqrt{11}$

4 $(x+2)^2-21$ $x=-2\pm\sqrt{21}$

5 a) $x=11.93, -2.933$ **b)** $x=-5, 2$
c) $x=1.046, -1.846$
d) $x=-58.25, -25.75$

CHAPTER 3
Exercise 4

1 $x=1.195, -6.195$

2 $x=5.179, 0.8206$

3 $x=1.530, -4.655$

4 $x=4.651, -2.251$

5 $x=2\pm\sqrt{\frac{19}{2}}$ **6** $x=1.2\pm\sqrt{1.4}$

7 $x=-2\pm\sqrt{34}$ **8** $x=-14\pm\sqrt{228}$

CHAPTER 3
Exercise 5

1 −0.382; −2.62 **2** 0.618; −1.62

3 2.29; −0.291 **4** 2.86; −0.698

5 2.37; 0.688 **6** $\frac{2\pm\sqrt{20}}{2}$

7 $\frac{4\pm\sqrt{8}}{4}$ **8** $\frac{-8\pm\sqrt{172}}{6}$

9 $x=\frac{3\pm\sqrt{57}}{6}$ **10** $x=\frac{5\pm\sqrt{73}}{4}$

CHAPTER 3
Exercise 6

1 $w=8$

2 $h=8$ Perimeter = 40 cm

3 $x=4$ or 12 Rectangle is 4 by 12

4 $4w(w+1)=35$ $w=2.5$ $65\frac{1}{2}$ m^2

5 $y=9.48$ 44.4 units^2

6 a) $20x^2+2x$ **b)** $x=1.53$ m

7 17.5 or 7.5

8 1.562 cm

9 $x=6.19$ cm

10 2.86 m so radii are 2.86 and 4.86

CHAPTER 3
Revision Exercise

1 a) $x=\pm4$ **b)** $x=4, -\frac{1}{2}$
c) $x=3, -2$ **d)** $x=4, -1$

2 $x=-11$ or 8

3 a) $(2x+7)(x-1)$ **b)** $x=-\frac{7}{2}$ or 1

4 $c=-7$ $x=\frac{1}{3}$

5 $w=\frac{3}{2}$

6 a) 4.760, −1.260 **b)** 4.513; −1.846

7 $(x+5)^2-32$ $x=-5\pm\sqrt{32}$

8 $2h^2+3h-100=0$ 6.36 cm

9 $r=8.48$ m Perimeter = 103.3 m

CHAPTER 4
Exercise 1

1 a) B(5, 9) **b)** D(−4, 7) **c)** $\overrightarrow{\mathbf{BD}}=\begin{pmatrix}-9\\-2\end{pmatrix}$

2 a) F(5, −1) **b)** G(7, 3) **c)** $\overrightarrow{\mathbf{EG}}=\begin{pmatrix}9\\-1\end{pmatrix}$
$\overrightarrow{\mathbf{EG}}=\overrightarrow{\mathbf{EF}}+\overrightarrow{\mathbf{FG}}$

3 a) $\overrightarrow{\mathbf{CD}}=\begin{pmatrix}6\\-12\end{pmatrix}$ **b)** $CD=\sqrt{180}$
c) M(9, 2)

4 a) $\overrightarrow{\mathbf{PQ}}=\begin{pmatrix}5\\12\end{pmatrix}$ **b)** $PQ=13$
c) M(−0.5, 2)

5 $\vec{UV} = \begin{pmatrix} 6 \\ -15 \end{pmatrix}$ $\begin{pmatrix} 2 \\ -5 \end{pmatrix}$ W(9, −2)

6 **a)** $\sqrt{73}$ **b)** $\sqrt{34}$ **c)** 12 **d)** $5a$

7 **a)** (4, 7) **b)** $(-\frac{3}{2}, \frac{17}{2})$
c) (−6, 2) **d)** $(-5p, 4p)$

8 PQ = $\sqrt{65}$ QR = $\sqrt{10}$ PR = $\sqrt{65}$
isos △

9 **a)** AB = 10 **b)** M(5, 2) N(8, −2)
c) MN = 5

10 **a)** K(9.5, 3.5) L(8.5, 0) M(6, 5.5)
b) $\vec{AK} = \begin{pmatrix} 4.5 \\ +1.5 \end{pmatrix}$ $\vec{AG} = \begin{pmatrix} 3 \\ +1 \end{pmatrix}$ G(8, 3)
c) G is on the line joining B to L
d) G is on the line joining C to M

CHAPTER 4
Exercise 2

1 **a)** $\begin{pmatrix} 2 \\ 9 \end{pmatrix}$ 4.5 **b)** $\begin{pmatrix} -4 \\ 8 \end{pmatrix}$ −2 **c)** $\begin{pmatrix} 10 \\ -2 \end{pmatrix}$ $\frac{-1}{5}$

2 **a)** 4 **b)** 3.6 **c)** $\frac{-1}{2}$
d) −1 **e)** $\frac{-1}{2}$ **f)** 3
g) −3 **h)** $\frac{-1}{3}$ **i)** −1

3 **a)** **ii)** parallel
iii) grad AB = 2 = grad CD
b) **ii)** parallel
iii) grad EF = $-\frac{1}{3}$ = grad GH
Parallel lines have the same gradient.

4 **a)** **ii)** perpendicular
iii) grad AB = 2 grad CD = $\frac{-1}{2}$
b) **ii)** perpendicular
iii) grad EF = −3 grad GH = $\frac{1}{3}$
Perpendicular lines have gradients that multiply to −1.

CHAPTER 4
Exercise 3

1 **a)** grad = 5, y-int = 9
b) grad = −4, y-int = 11
c) grad = $-\frac{1}{2}$, y-int = 7
d) grad = 2, y-int = −7
e) grad = −3, y-int = −9

2 **a)** $y = 6x + 14$ **b)** $y = 2x - 5$
c) $y = \frac{-1}{4}x + 1$ **d)** $y = 12x - 11$

3 **a)** $y = 4x - 5$ **b)** $y = 2x + 9$
c) $y = -3x + 1$ **d)** $y = -2x - 9$

4 3 $y = 3x - 13$

5 **a)** $y = 3x - 1$ $y = \frac{-1}{2}x + 7$

CHAPTER 4
Exercise 4

1 **a)** $y = 3x + 5$ **b)** $3y - 4x + 14 = 0$
c) $4y + 3x - 22 = 0$ **d)** $7y + 4x = 62$

2 **a)** grad = $\frac{-3}{4}$ y intercept = 3
b) −2 6
c) 2 5
d) $\frac{5}{2}$ 4
e) $\frac{-2}{7}$ 2
f) $\frac{5}{2}$ −3
g) $\frac{3}{2}$ 3
h) −4 7
i) $\frac{3}{4}$ $\frac{3}{2}$

CHAPTER 4
Exercise 5

1 −1 −1 parallel

2 $\frac{-1}{2}$ 2 perpendicular

3 3 $\frac{-1}{3}$ perpendicular

4 AB = 10
AC = 5
Area = 25

5 L_1, L_4 are //
L_3, L_5 are //
L_2, L_3 are ⊥
L_2, L_5 are ⊥
L_1, L_6 are ⊥
L_4, L_6 are ⊥

6 **a)** $y = 2x - 8$ **b)** $4y + x + 8 = 0$
c) $y = -4x + 14$ **d)** $4y - x = 0$
e) $y = \frac{1}{2}x + \frac{9}{2}$

7 $5x - y = 10$

8 $y = \frac{3}{4}x - \frac{1}{2}$

9 **a)** A(0, −6) B(3, 0)
b) $y = 2x - 16$ (0, −16)
c) $y = \frac{-1}{2}x + 6$ (12, 0)

CHAPTER 4

Exercise 6

1 (4, –1) **2** (4, 3) **3** (5, –2)

4 (2, 1) **5** (2, –1) **6** (2, 5)

7 (7, 2) **8** (1, 5) **9** (5, 2)

10 (3, 5)

11 **b)** $y = 2x - 11$ **c)** $y = -3x + 14$
d) $x = 5, y = -1$
e) Rotation of 90° anticlockwise about (5, –1)

12 **a)** $2x + y = 9$ **b)** $3y - 4x = -3$
c) P(3, 3)

CHAPTER 4

Exercise 7

1 **a)** 10 **b)** 18 **c)** $33\frac{1}{2}$

2 24

3 **i)** $y = \frac{-2}{3}x - 2$
D(–3, 0)
ii) C(1, 6) **iii)** Area = 26

4 **i)** $3x + 4y = 5$ **ii)** B(3, –1)
iii) AD is $y = x + 3$
CD is $3x + 4y = 19$
iv) D(1, 4) **v)** Area = 14

5 **a)** R(13, –2) **b)** $y = \frac{1}{2}x - \frac{7}{2}$
c) Q(3, –2)
S(19, 6)
d) 80 units^2

6 **a)** $y = \frac{-2}{3}x + 8$ **b)** A(12, 0) B(0, 8)
c) Area = 26

7 **ii)** $y = \frac{1}{2}x + 4$
iv) L is the angle bisector on ∠BAC

CHAPTER 4

Revision Exercise

1 $\sqrt{40}$
$\sqrt{40}$ ABC is isosceles

2 $\frac{2}{5}$
$2y + 5x = 22$

3 **a)** **i)** (2, 2) **ii)** (–8, 4) **b)** T(–3, 6)

4 grad = –2.5
$2y + 5x - 23 = 0$
$5y - 2x - 14 = 0$

5 $y = -5x + 33$
$y = \frac{1}{5}x + \frac{9}{5}$
13 units^2

6 grad = $\frac{-1}{4}$
$x = 1, y = 2$
$y - 4x + 2 = 0$

7 A(–1, –2)
C(7, 8)
B(6, –1)
D(0, 7)

8 $y = \frac{1}{2}x + \frac{3}{2}$
$y = \frac{1}{7}x + \frac{13}{7}$
CP = 5 CQ = 5 CR = 5
Circle passes through P, Q and R

9 **a)** $y = 4x - 5$ **b)** M(3, 7)
c) C(4, 11)

10 $y = 5x - 6$

11 $y = \frac{1}{2}x - 4$
$x = 2, y = -3$
$\sqrt{80}$

12 **i)** $x + 3y - 3 = 0$
iv) $\sqrt{90}$

CHAPTER 5

Exercise 1

1 **a)** $5\sqrt{3}$ **b)** $3\sqrt{6}$ **c)** $3\sqrt{2}$
d) $6\sqrt{2}$ **e)** $40\sqrt{2}$ **f)** $\frac{3}{5}\sqrt{2}$
g) $\frac{5}{11}\sqrt{3}$ **h)** $\frac{8\sqrt{3}}{15}$ **i)** $\frac{2}{3}\sqrt{10}$
j) $\frac{5}{3}$ **k)** $\frac{4}{\sqrt{3}} = \frac{4\sqrt{3}}{3}$ **l)** $4\sqrt{15}$

2 **a)** $5\sqrt{2}$ **b)** $2\sqrt{3}$ **c)** $8\sqrt{5}$
d) $3\sqrt{2} + 3\sqrt{3}$ **e)** $2\sqrt{7} - 2\sqrt{3}$

3 **a)** $10 - \sqrt{5}$ **b)** $-5 + 7\sqrt{5}$

4 **a)** $12 + \sqrt{3}$ **b)** $16 + 5\sqrt{2}$
c) $1 + \sqrt{5}$ **d)** $16\sqrt{2} - 11$
e) 2 **f)** 37
g) $8 + 2\sqrt{7}$ **h)** $101 - 36\sqrt{5}$

5 **a)** $\frac{3\sqrt{7}}{7}$ **b)** $\frac{\sqrt{10}}{5}$ **c)** 2
d) $\frac{\sqrt{10}}{2}$ **e)** $\frac{\sqrt{5}}{5}$ **f)** $5\sqrt{2}$
g) $4\sqrt{3}$ **h)** $\frac{16}{7}$

6 **a)** $2 \pm 5\sqrt{3}$ **b)** $-3 + 3\sqrt{2}$

7 $4\sqrt{3}$

8 **a)** 17 **b)** 6 **c)** 29 **d)** $a^2 - b^2x^2$

CHAPTER 5
Exercise 2

1 **a)** $\frac{24 - 3\sqrt{2}}{62}$ **b)** $\frac{64 + 25\sqrt{7}}{93}$
c) $\frac{23 - 16\sqrt{2}}{17}$

2 **a)** $\frac{27 - 9\sqrt{2}}{7}$ **b)** $\frac{75 + 74\sqrt{5}}{95}$
c) $17 - 9\sqrt{3}$ **d)** $\frac{116 + 38\sqrt{7}}{93}$
e) $\frac{17 + 10\sqrt{10}}{79}$

3 $10 - 3\sqrt{3}$

6 **a)** $5\sqrt{7}$ **b)** $\frac{6\sqrt{10}}{5}$ **c)** $5 + 22\sqrt{2}$
d) $48 - 24\sqrt{3}$ **e)** $\frac{21 + 11\sqrt{5}}{4}$

7 **a)** $w = 7 + \sqrt{5}$ **b)** $-2 + 3\sqrt{3}$
c) $5 - 2\sqrt{7}$

8 **a)** $p = 2\sqrt{3}$ $q = 2\sqrt{2}$
b) $p = 1 + \sqrt{3}$ $q = 4 - \sqrt{3}$

9 **a)** $-2\sqrt{5}, 6\sqrt{5},$ **b)** $\frac{+1}{2}\sqrt{3}, -4\sqrt{3}$

CHAPTER 5
Revision Exercise

1 **a)** $2\sqrt{2}$ **b)** $-34 + 22\sqrt{5}$
c) $16 - 6\sqrt{7}$ **d)** $\frac{17 + 7\sqrt{5}}{11}$

2 **a)** $k = 11$ **b)** $k = 17$

3 $2\sqrt{3}$ cm

4 $x = 11\sqrt{2}$

5 $p = \frac{12}{\sqrt{3}} = 4\sqrt{3}$ $q = -3\sqrt{3}$

6 $x = -4 \pm 3\sqrt{5}$

7 **a)** $14 - 3\sqrt{3}$ **b)** $8 - 3\sqrt{7}$
c) $3 + 5\sqrt{2}$

9 $x = 7\sqrt{3}$ or $3\sqrt{3}$

CHAPTER 6
Exercise 1

13 $p = -5$ $q = 3$ $k = 2$ $(-1, -32)$

14 $c = -4$ $d = 8,$ $k = \frac{1}{2}$ $(0, 16)$

15 $s = -5$ $t = 10$ $k = \frac{1}{5}$ $(0, 10)$

CHAPTER 6
Exercise 2

1 $3(x + 2.5)^2 - 18.75$

2 $4(x + 3)^2 - 34$

3 $3(x - 2)^2 + 3$ **4** $6(x - 2.5)^2 - 46.5$

5 $2(x + \frac{5}{4})^2 - \frac{17}{8}$ **6** $4(x - \frac{3}{4})^2 - \frac{33}{4}$

7 $4(x + \frac{9}{8})^2 - \frac{33}{16}$ **8** $10(x - 0.3)^2 - 7.9$

9 $-2(x - \frac{3}{2})^2 + \frac{23}{2}$ **10** $-5(x - 0.9)^2 + 2.05$

11 $-3(x - 1.5)^2 + 4.75$

12 $5(x - 0.8)^2 + 3.8$

CHAPTER 6
Exercise 3

1 **a)** $y = (x - 2)^2 - 11$
b) $y = (x - 6)^2 + 11$

2 **a)** $y = 3(x - 4)^2 - 43$
b) $y = 5(x - 4)^2 - 66$
c) $y = -3(x + 3)^2 + 47$
d) $y = -0.5(x - 6)^2 + 19$

3 $3(x - 1)^2 + 9$ $x = 1$ $(1, 9)$

4 $5(x + 0.8)^2 - 7.2$ $x = -0.8$

5 **a)** $q = -2$ $r = -25$
b) A(−3, 0) B(7, 0)

CHAPTER 6
Exercise 4

1 a) $x < 7$ **b)** $x > -7$ **c)** $x > 3$

2 a) $2 < x < 8$ **b)** $-4 < x < 6$

3 a) $(4, 7)$ **b)** $x < 4$

4 a) $-12 < x < 4$ **b)** $-2 < y < 10$
c) $-\frac{1}{3} < y < 3$

5 $x < -2 - \sqrt{15}$ $x > -2 + \sqrt{15}$

6 a) $x \leqslant 0$ or $x \geqslant 6$
b) $-2 \leqslant x \leqslant 8$
c) $-1 < y < 5$

7 a) $x \leqslant -8$ or $x \geqslant 2$ **b)** $2 \leqslant x \leqslant 4$

8 a) $-\frac{5}{2} \leqslant x \leqslant -1$
b) $2x^2 + 7x + 7 = 2(x + \frac{7}{4})^2 + \frac{7}{8}$
so never negative

9 $w(16 - w) > 48$ $4 < w < 12$

10 $r < 2$ or $r > 7$

11 $0 \leqslant 2x^2 + 4x - 96$ $x \geqslant 6$

CHAPTER 6
Exercise 5

1 a) 2 **b)** 1 **c)** 0 **d)** 2

2 $k < \frac{49}{8}$

3 $k > 12$ or $k < -12$

4 $-24 < k < 24$

5 $p > 2$ or $p < -\frac{2}{9}$

6 $p < 0$ or $p > 4$

CHAPTER 6
Revision Exercise

2 a) i) -20
ii) no solutions
Graph is ∪ shaped so always ⊕.
b) $k = -\frac{1}{2}$

3 a) $p = 2$ $q = 1.5$ $r = -11.5$
b) $x = -1.5$
c) $(-1.5, -11.5)$

4 a) $x \leqslant -\frac{15}{4}$ or $x \geqslant 2$
b) $3 - \sqrt{17} \leqslant x \leqslant 3 + \sqrt{17}$

5 $m < -\frac{3}{11}$ or $m > 3$

6 a) i) $a = -\sqrt{2}$ $b = 2$ **ii)** $x = \sqrt{2}$
b) $k < -9$ or $k > 7$

7 $x < \dfrac{5 - \sqrt{19}}{2}$ or $x > \dfrac{5 + \sqrt{19}}{2}$

8 $s = 4$ $t = -5$ $u = -18$ $(\frac{5}{4}, 18)$

9 $x > 2\sqrt{5} + 6$ or $x < 2\sqrt{5} - 6$

10 $p = -3$ $q = 2$ $r = 32$
$-2 - \sqrt{\frac{32}{3}} \leqslant x \leqslant -2 + \sqrt{\frac{32}{3}}$

11 $a = 2$ $p = 1$ $q = -\frac{1}{2}$

12 $k < -4.8$

CHAPTER 7
Exercise 1

1

x	1	2	3	−1	−2	−3
Grad	2	4	6	−2	−4	−6

Gradient = $2a$

CHAPTER 7
Exercise 2

1 a) Grad PQ are 12.1, 12.01, 12.001, 12.0001

2

P	1	2	3	4	5	6
Grad PQ	2.001	4.001	6.001	8.001	10.001	12.001
Est of grad	2	4	6	8	10	12

Gradient of $y = x^2$ at $(9, 9^2) = 2a$

4 a) Grad est = 17
b) Grad est = 16
c) Grad PQ = 9.893
d) Grad est = −6

CHAPTER 7
Exercise 3

1 a)

x	0	1	2	3	4	5	6
Grad est	0	3	12	27	48	75	108

b) Gradient $= 3x^2$

2 a)

x	0	1	2	3	4	5
Grad est	0	4	32	108	216	500

b) Gradient $= 4x^3$

3 Gradient $= 5x^4$

4 a)

x	1	4	9	16	25	100
Grad est	0.5	0.25	0.16	0.125	0.1	0.05
	$\frac{1}{2}$	$\frac{1}{4}$	$\frac{1}{6}$	$\frac{1}{8}$	$\frac{1}{10}$	$\frac{1}{20}$

b) Gradient $= \dfrac{1}{2} \times \dfrac{1}{\sqrt{x}} = \frac{1}{2}x^{-\frac{1}{2}}$

5 a)

x	1	2	3	4	5	6
Grad est	−1	−0.25	−0.111	−0.625	−0.04	−0.0277
		$-\frac{1}{4}$	$-\frac{1}{9}$	$-\frac{1}{16}$	$-\frac{1}{25}$	$-\frac{1}{36}$

b) Gradient $= -\dfrac{1}{x^2} = -x^{-2}$

6 Gradient $= -\dfrac{2}{x^3}$

7

Curve	x^2	x^3	x^4	x^5	$x^{\frac{1}{2}}$	x^{-1}	x^{-2}
Grad	$2x$	$3x^2$	$4x^3$	$5x^4$	$\frac{1}{2}x^{-\frac{1}{2}}$	$-x^{-2}$	$-2x^{-3}$

Gradient of $y = x^n$ is nx^{n-1}

CHAPTER 7
Exercise 4

1 Gradient = 8

2 $\dfrac{dy}{dn} = 4.5x^{3.5}$

3 $\dfrac{dy}{dt} = -3t^{-4} = -\dfrac{3}{t^4}$

4 $f'(-2) = 80$

5 $\frac{1}{8}$

6 $\frac{1}{12}$

7 $6p^5$

8 a) $x^{-\frac{3}{2}}$ **b)** $\frac{-3}{64}$

9 a) $x^{-\frac{5}{2}}$ **b)** $-\frac{5}{256}$

10 a) $x^{\frac{5}{3}}$ **b)** $\frac{5}{3}x^{\frac{2}{3}}$

11 a) $x^{\frac{1}{2}}$ **b)** $\frac{1}{6}$

12 a) $t^{\frac{5}{3}}$ **b)** $\frac{20}{3}$

CHAPTER 7
Exercise 5

1

x	0	1	2	3	4	5
Grad est	0	8	28	60	104	150

$\dfrac{dy}{dx} = 6x^2 + 2x$

2

x	0	1	2	3	4	5
Grad est	7	17	27	37	47	57

$\dfrac{dy}{dx} = 10x + 7$

3

x	0	1	2	3	4	5
Grad est	0	5	16	33	56	85

$\dfrac{dy}{dx} = 3x^2 + 2x$

4

x	0	1	2	3	4	5
Grad est	−5	−1	3	7	11	15

$\dfrac{dy}{dx} = 4x - 5$

5

x	1	2	3	4	5
Grad est	−31	−4	1	2.75	3.56

$\dfrac{dy}{dx} = 5 - \dfrac{36}{x^2}$

6 $\dfrac{dy}{dx} = anx^{n-1} + bmx^{m-1}$

CHAPTER 7
Exercise 6

1 **a)** $3x^2 + 3$ **b)** $6x + 5$
c) $12x^2 - 10x$ **d)** $18x^2 + 6x - 2$
e) $-\dfrac{24}{x^3}$ **f)** $\dfrac{4}{\sqrt{x}}$
g) $10x - \dfrac{7}{x^2}$ **h)** $\dfrac{-3}{x^2} + \dfrac{10}{x^3}$
i) $1.5x^{-0.5} + 2x^{-1}$

2 **a)** $12x - 8$ **b)** $3x^2 + 2x - 9$
c) $3.6x^{0.2} - 1.2x^{-0.4}$ **d)** $-4x^{-2} - 3x^{-4}$

3 **a)** $6x + 2 + 5x^{-2}$ **b)** $8x^3 - 3x^2 + 10$

4 **a)** $6t + 6t^{-2}$ **b)** $\dfrac{dp}{dt} = 1 + 5t^{-2}$

5 **a)** 25 **b)** $-\frac{1}{2}$ **c)** 2 **d)** 48

6 $(\sqrt{2}, 7\sqrt{2})$ $(-\sqrt{2}, -7\sqrt{2})$

7 (1, 9) (2, 8)

8 (1, −4) $(\frac{7}{3}, \frac{-392}{27})$

9 (10, 0.4) (−10, −0.4)

10 **a)** $\frac{-5}{3}x^{-2} + \frac{14}{3}x^{-3}$
b) $-3x^{-2} + 2x^{-3} + 9x^{-4}$
c) $\frac{1}{2} - 2x^{-2}$

CHAPTER 7
Revision Exercise

1 **a)** $28x^3 - 16x$ **b)** $100x^3 - 20x$
c) $16x^3 - 9x^2 + 16x - 10$
d) $2.5x^{-\frac{1}{2}} + 14x$ **e)** $-8x^{-\frac{3}{2}}$
f) $0.2x^{-0.8}$ **g)** $1 - 5x^{-2}$
h) $2x - 18x^{-3}$ **i)** $\frac{5}{3} - 3x^{-2}$
j) $\frac{-5}{2}x^{-2} + 7x^{-3}$

2 **a)** −8 **b)** 2 **c)** 36
d) −4 **e)** $\frac{-13}{9}$ **f)** $\frac{12}{125}$

3 (2, 4); (−4, 28)

4 Point (4, 16)

5 Point (2, 15)

6 $(8, \frac{1}{2})$

7 **a)** $12t^3 - 6t^2$ **b)** $-4u^{-2} + 5$

8 (2, 4) or (−2, −4)

9 $k = -10$

10 56

11 14

12 $a = 12$ $b = -4$

CHAPTER 8
Exercise 1

1 $u = 2, -1$ 2 $t = \pm\frac{1}{2}; \pm 3$

3 $x = 4$ or 9 4 $t = \pm 2$ or ± 3

5 $q = \frac{-1}{2}, 3$ 6 $x = 2, \frac{-1}{8}$

7 $x = 27$ or −8 8 $x = \frac{1}{16}$ or 1

9 $x = 0$ or 1 10 $x = \frac{1}{2}$ or 1

11 $x = 4$ 12 $x = -1$ or 4

13 $x = -3$ or 2 14 $y = 4$ or $\frac{16}{9}$

CHAPTER 8
Exercise 2

1 $x = 2, y = 4$ $x = 5, y = 25$

2 $x = 9, y = -6$ $x = 1, y = 2$

3 $x = 1, y = 3$ $x = 3, y = 1$

4 $x = 8, y = 35$ $x = -1, y = -1$

5 $p = 4, q = -1$

6 $x = 2, y = 3$ $x = 5, y = 36$

7 $x = 2, y = 8$

8 $q = 2, p = -1$ $q = -5, p = 13$

9 $x = 5, y = 3$ $x = \frac{17}{5}, y = \frac{-1}{5}$

10 $y = -2, x = 2$

11 $x = 4, y = -4$ $x = 1, y = 2$

12 (3, 6) or (5, −2)

CHAPTER 8
Exercise 3

1 **ii)** 2

2 1
Line is tangential to curve

3 a) (7, 35) (−1, 11)

5 $c = -17$

CHAPTER 8
Revision Exercise

1 a) $y = \pm 2$ **b)** $t = 1$ or -2
c) $p = 27$ or $\frac{1}{8}$

2 (−2, −11) (4, 43)

3 a) i) −131 **ii)** No solutions

4 $x = 25$

5 $x = \pm 2$ or $\pm\sqrt{6}$

6 $x = 9$ or 16

7 $x = 2, y = 6$ $x = \frac{-4}{3}, y = \frac{-2}{3}$

8 a) $x = 4, y = 10$ $x = -5, y = -8$
c) $k = \frac{-73}{4}$

9 (4, 7) and (−2, −17) $-2 < x < 4$

CHAPTER 9
Exercise 1

1 2.804

2 7.42

3 a) $\pi + 0.4\pi t$ **b)** 0.4π
c) $0.25\pi + 0.2\pi t + 0.04\pi t^2$

4 $x < -4$

5 $x > 9$

6 $x < -2$ or $x > \frac{2}{3}$

7 $\frac{1}{2} < x < \frac{2}{3}$

CHAPTER 9
Exercise 2

1 (3, −9) Min

2 (1, 2) Max (−1, −2) Min

3 Max (−2, 16) Min (2, −16)

4 Max (−2, 31) Min (4, −77)

5 Max (−3, 81) Min (0, 0)
Max (3, 81)

6 a) $x = \pm 1, \pm 7$
b) Min (−5, −576) Max (0, 49)
Min (5, −576)
d) i) 4 **ii)** 4 **iii)** 0

7 Max (2, 20) Min (4, 16) $16 < k < 20$

8 a) can't divide by 0
b) y is large negative number
c) y is large positive number
d) (2, 12) is min

9 a) can't divide by 0
b) y is large negative
c) y is large negative
d) Max (−3, −6); Min (3, 6)

10 Max $(4, \frac{9}{8})$

11 Max (0, 0) Min (243, −729)

CHAPTER 9
Exercise 3

1 $x = 150$ m, L = 300 m

2 $r = 4.12$ cm $h = 8.24$ cm
Minimising S will minimise amount of metal to be used
ignored lips of can; is can holdable?

3 c) $\frac{dP}{dr} = \frac{-200}{r^2} + 2 + \frac{1}{2}\pi$ $r = 7.484$ m

4 $x = 20$

5 a) $n = -45s + 10\,000$
b) $C = 60\,000 + 50N$
d) $s = £136.11$ $N = 3875$ bikes

6 $v = 32$ km/hr

CHAPTER 9
Exercise 4

1 a) $y = 4x - 4$ $y = -\frac{1}{4}x + \frac{9}{2}$
b) $y = 2x + 3$ $y = -\frac{1}{2}x + \frac{11}{2}$

3 $y = 9x - 27$ $y = -\frac{1}{9}x - \frac{1}{3}$

5 $y = x + 3$ $y = \frac{1}{2}x + 6$ $x = 6, y = 9$

6 (0, 2) Max (2, −2) Min
$y = 9x - 25$ (0, −25)

7 $y = \frac{3}{2}x - 10$ B(−1, −0.7)

8 $a = 2$ $k = -26$

9 $y = \frac{-1}{2p}x + p^2 + \frac{1}{2}$ M(0, $\frac{1}{4}$)

10 $3y + 2x - 76 = 0$

CHAPTER 9

Exercise 5

1 a) $80x^3 + 14$ **b)** $36x^2 + \frac{10}{x^3}$
c) $-1.25x^{-1.5}$ **d)** $30x^4 + 20x^{-6}$
e) $6x^{-3} - 12x^{-4}$

2 a) $12t + 30t^{-4}$ **b)** $g''(4) = -\frac{3}{8}$

3 a) SP (4, −16) Min, value of $\frac{d^2y}{dx^2} = 2$
b) SP (5, 25) Max, value of $\frac{d^2y}{dx^2} = -2$
c) SP (0, 0) Max, value of $\frac{d^2y}{dx^2} = -6$
(2, −4) Min, value of $\frac{d^2y}{dx^2} = 6$
d) SP (0, 7) Max, value of $\frac{d^2y}{dx^2} = -16$
(2, −9) Min value of $\frac{d^2y}{dx^2} = 32$
(−2, −9) Min value of $\frac{d^2y}{dx^2} = 32$

$\frac{d^2y}{dx^2} > 0 \Rightarrow$ SP is Min
$\frac{d^2y}{dx^2} < 0 \Rightarrow$ SP is Max

CHAPTER 9

Exercise 6

1 a) $x = 2, -14$ **b)** SP (−6, −64)
c) $\frac{d^2y}{dx^2} = 12$ at SP $\Rightarrow$ Min

2 a) $x = 1, 9$ **b)** SP (3, −4); (−3, −16)
c) when $x = 3$ $\frac{d^2y}{dx^2} = \frac{2}{3} \Rightarrow$ Min
$x = 3$ $\frac{d^2y}{dx^2} = -\frac{2}{3} \Rightarrow$ Max

3 (0, 0) Min
(2, 16) Max
(−2, 16) Max
$0 < k < 16$

4 (2, 27) Min (−2, −37) Max

CHAPTER 9

Revision Exercise

1 a) $3x^2 - 12x + 9$
b) (1, 6) Max (3, 2) Min

2 b) (2, −27) Min (−1, 0) Max
d) $k > 0$ or $k < -27$

3 a) $3x^2 - 6x$ **b)** $y = 24x - 80$
c) $a = 6$ $b = -25$

4 a) $x + 6y = 57$ **b)** Q($-\frac{19}{6}$, $\frac{361}{6}$)

5 $x = 7.37$ cm

6 b) $12x^2 - 36$
d) Max (0, 90.25) Min (3, 9.25)
Min (−3, 9.25)
e) (3, 9) (−3, 9)

7 c) Q is (−4, −77)

8 c) $r = 10$ 114.6°

9 $-\frac{1728}{125}$ or 8, $16y + x - 1288 = 0$

10 $x = \frac{81}{16}$ SP is a minimum

CHAPTER 10

Exercise 1

1 a) $(x-4)^2 + (y-2)^2 = 25$
b) $(x-6)^2 + (y+4)^2 = 49$
c) $(x-3)^2 + y^2 = 36$
d) $(x+2)^2 + (y+3)^2 = 36$

2 a) AB = 10 **b)** (−1, 3)
c) $(x+1)^2 + (y-3)^2 = 25$

3 $(x+4)^2 + (y-11)^2 = 169$

4 a) $(x-4)^2 + (y-1)^2 = 25$
b) $(x+3)^2 + (y-8)^2 = 289$

5 a) C(8, 3) $r = 7$ **b)** C(3, 2) $r = 4$
c) C(−5, 2) $r = 2$ **d)** C(−4, −1) $r = \sqrt{7}$

6 a) C(−2, 3) $r = 5$ **c)** grad CP = $\frac{3}{4}$
d) gradient = $-\frac{4}{3}$ **e)** $3y + 4x = 26$

7 $y = -x + 6$

8 **a)** $(x+6)^2 - 36$ **b)** $(y-4)^2 - 16$
c) $C(-6, 4)$ $r = 2$

9 $C(-9, -5)$ $r = 10$

10 $C(4, 2)$ $r = 3$

11 centre $(-f, -g)$ radius $\sqrt{f^2 + g^2 - c}$
point No curve

CHAPTER 10
Exercise 2

1 $(x-3)^2 + (y+4)^2 = 169$
$y = \frac{-5}{12}x + \frac{136}{12}$ $y = \frac{12}{5}x + \frac{113}{5}$

2 $C_1(-3, 2)$ $r = 5$ $y = \frac{-4}{5}x + \frac{19}{3}$
$(x+3)^2 + (y-9)^2 = 4$ $B(-3, 7)$
$y = 7$ $(\frac{-1}{2}, 7)$

3 $C_1(4, 2)$ $r_1 = 5$ $C_2(-5, -10)$ $r_2 = 10$
$(x+2)^2 + (y+6)^2 = 225$

4 $(x-3)^2 + (y+2)^2 = 25$

5 $C(3, -4)$ $r = 2$
Min value of AP $\sqrt{20} - 2$
Max value of AP $\sqrt{20} + 2$

6 **a)** 5 **b)** $C(5, 0)$ **c)** $4y + 3x = 15$

CHAPTER 10
Exercise 3

1 **a)** $x^2 + y^2 = 100$
Points (6, 8) or (8, −6)

2 Points (6, −3) and (8, 1)

3 **a)** Points (4, 1) *or* (0, 3)

4 **b)** $k = \pm 2\sqrt{5}$

5 **a)** $x^2 + y^2 = 4$ **b)** $y = mx - 4$
d) **i)** two points of intersection
ii) one point of intersection
iii) no points of intersection

6 $(x-1)^2 + (y-2)^2 = r^2$

7 $x = 3$, $y = 4$
line is tangent to circle at (3, 4)

CHAPTER 10
Revision Exercise

1 **b)** $(x-2)^2 + (y+3)^2 = 65$
$x^2 + y^2 - 4x + 6y - 52 = 0$
c) $y = \frac{4}{7}x + \frac{36}{7}$

2 **a)** $C(1, -3)$ $r = 5$ **b)** $5\sqrt{2}$

3 **b)** $x = 3 \pm \sqrt{5}$

4 **a)** $r = 10$ **b)** $p = -4$ or -8

5 **a)** $y = \frac{1}{2}x + \frac{13}{2}$ **b)** $C(-13, 0)$ $\sqrt{505}$

7 $(x-3)^2 + (y - 3\frac{1}{2})^2 = 6.25$

8 C_1 has centre $P_1(0, 0)$ radius $\sqrt{20}$
C_2 has centre $P_2(12, 6)$ radius $\sqrt{80}$
P(4, 2)

9 **a)** $(x-3)^2 + y^2 = 9$ **b)** $y = mx + 2m$
d) $m = \pm\frac{3}{4}$

CHAPTER 11
Exercise 1

7 $y = \dfrac{18}{x}$ 8 $y = -5x^3$ 9 $y = 20\sqrt{x}$

CHAPTER 11
Exercise 2

7 $y = 2(x+2)(x-4)$

8 $y = \frac{1}{2}(x+4)(x+1)(x-5)$

9 $y = 2(x+2)(x-3)^2$

10 $4x^2(x-3)^2$

11 $(x-5)(2x-3)(x-2)$

12 $(x-2)(x+7)(x-5)$

CHAPTER 11
Exercise 3

1 **a)** reflection in x axis
b) reflection in x axis
c) reflection in x axis

2 **a)** $\begin{pmatrix}0\\7\end{pmatrix}$ **b)** $\begin{pmatrix}0\\-4\end{pmatrix}$ **c)** $\begin{pmatrix}0\\5\end{pmatrix}$

3 a) $\begin{pmatrix}2\\0\end{pmatrix}$ **b)** $\begin{pmatrix}-2\\0\end{pmatrix}$ **c)** $\begin{pmatrix}3\\0\end{pmatrix}$

4 a) stretch ×3 in y direction
b) stretch ×2 in y direction
c) stretch ×4 in y direction

CHAPTER 11
Exercise 4

1 $y = \frac{-12}{x}$

2 a) $y = x^2 + 4$ **b)** $y = \sqrt{x-4}$
c) $y = \frac{1}{x} - 5$ **d)** $y = (x+2)^3$
e) $y = \sqrt{x} + 2$ **f)** $y = \sqrt{x-5}$

3 a) translation $\begin{pmatrix}-7\\0\end{pmatrix}$
b) translation $\begin{pmatrix}0\\7\end{pmatrix}$
c) reflection in x axis
d) translation $\begin{pmatrix}3\\0\end{pmatrix}$
e) translation $\begin{pmatrix}0\\10\end{pmatrix}$
f) reflection in x axis
g) translation $\begin{pmatrix}10\\0\end{pmatrix}$
h) translation $\begin{pmatrix}4\\2\end{pmatrix}$

4 b) $y = 4x - 3$ **c)** translation $\begin{pmatrix}4\\0\end{pmatrix}$
d) $y = 4x - 19$ **e)** $y = -4x + 13$

CHAPTER 11
Exercise 5

1 a) $y = 4(x^2+3) = 4x^2 + 12$
b) $y = \left(\frac{x}{2}\right)^2 + 3 = \frac{1}{4}x^2 + 3$
c) $y = (5x)^2 + 3 = 25x^2 + 3$

2 a) $y = 2(x-1)(x-4)$
b) $y = \left(\frac{x}{3} - 1\right)\left(\frac{x}{3} - 4\right)$
c) $y = (2x-1)(2x-4)$

3 a) $y = \frac{45}{x^2}$ **b)** $y = \frac{36}{x^2}$ **c)** $y = \frac{1}{x^2}$

6 a) A(2, 0) B(4, 0) C(0, 8) P(3, −1)

CHAPTER 11
Revision Exercise

1 b) $y = \frac{1}{2}x^3$

2 b) $y = 10\sqrt{x}$ **c)** $y = 2\sqrt{x-3}$
d) translation $\begin{pmatrix}0\\-7\end{pmatrix}$

5 b) $y = \frac{4}{x^2}$ **c)** Translation $\begin{pmatrix}-5\\0\end{pmatrix}$

6 $a = 1$, $b = -3$, $k = 5$, $p = 120$

9 +9
translation $\begin{pmatrix}-3\\0\end{pmatrix}$
9

REVISION 1
Indices and Surds

1 a) $\frac{1}{25}$ **b)** 4 **c)** 18 **2** $a^{-\frac{2}{3}}$

3 a) $8 + 2\sqrt{2}$ **b)** $3\sqrt{2}$ **c)** $6 + 2\sqrt{2}$

4 a) 10 **b)** 81 **5** $\frac{1}{3a^2}$ **6** $x = 2$

7 a) 1 **b)** 4 **c)** $\frac{1}{8}$ **8** $p = 9$ $q = 4$

9 a) 31 **b)** $\frac{125}{27}$ **10** $(6\sqrt{2}, 53)$

11 x^{-3} $(2, \frac{1}{8})$ **12** $\frac{15}{2} + 4\sqrt{3}$ **13** $6\sqrt{2}$

14 b) $x = -2$ or 2

15 $2\sqrt{5} + 5\sqrt{3}$

16 a) 200 **b)** 64 **c)** $\frac{1}{5}$ **d)** 10

REVISION 2
Polynomials

1 $x^3 + x^2 - 2$

2 $x = -\frac{5}{4}, y = \frac{11}{2}$ $x = 2, y = -1$

3 a) $x > 3$ **b)** $-3 \leqslant x < 6$

4 $x = 6 + \sqrt{41}$ or $6 - \sqrt{41}$

5 $(x+7)(x-1)$

6 $k < -12$ or $k > 12$

7 a) $\frac{37}{10}$ m **b)** $t = 8$

8 $x = 2, y = -1$ or $x = -1, y = 2$

9 a) $x = 7\sqrt{5}$ or $\sqrt{5}$
b) $x \leqslant \sqrt{5}$ or $x \geqslant 7\sqrt{5}$

10 $-3 < a < 3$

11 $1 \leqslant x \leqslant 12$

12 a) $(x+3)^2 - 22$
b) $x = -3 + \sqrt{22}$ or $-3 - \sqrt{22}$
c) $x < -3 - \sqrt{22}$ or $x > -3 + \sqrt{22}$

13 $x^3 + 4x^2 + 6$ $x^5 + 2x^4 + 5x^3 + 6x^2 - 20$
$4x^2 - 20x + 30$

14 $x = -1, y = -3$ $x = \frac{3}{2}, y = 2$

15 $(7x-2)(7x-1)$ $y = \frac{4}{49}$ or $\frac{1}{49}$

16 a) i) -116 **ii)** none
iii) always above x axis
b) $\frac{16}{5}$

17 $p = 5, q = 2, r = 10$

18 $a = 2, p = 2, q = -5.5$

19 a) $6x(x-3)(x+3)$ **b)** $2(5x+2)(x-3)$

20 $x < -1$ or $x > \frac{2}{3}$

21 $y = 2, x = 1$

22 i) $A(p, p-p^2)$ **ii)** $0 < p < 1$
iii) $p = \dfrac{-1+\sqrt{5}}{2}$

23 ii) $u = \frac{1}{8}$ or 27

24 $y = -3(x-2)^2 + 17$

REVISION 3

Co-ordinate geometry and graphs

1 $2\sqrt{10}$ Midpoint: (3, 0)

2 $y = \frac{1}{2}x + 2$

3 $y = 20x^3$ or $y = 5x^5$

4 $(x+1)^2 + (y-10)^2 = 25$

5 $2y + 3x - 43 = 0$ $3y - 2x - 16 = 0$

6 $-4 \leqslant x \leqslant \frac{5}{2}$ or $x \geqslant 6$

7 $x = 2, y = 5$
The line $y = 9x - 13$ is a tangent to $y = 2x^2 + x - 5$ at (2, 5)

8 A(8, 2) **9 i)** $k = 100, n = 2$

10 $2y + 5x - 11 = 0$

11 $C_1(5, -2)$ $r = 5$
$(x-14)^2 + (y-10)^2 = 100$ C(8, 2)
$4y + 3x - 32 = 0$

12 C(−4, 3) $r = 13$ $\sqrt{202} + 13$ AT $= \sqrt{33}$

13 i) (1, 5) **ii)** (−2, 5) **iii)** (2, 15)
iv) $(\frac{1}{2}, 5)$

14 i) $2y - x = -9$ **ii)** $y = -2x - 2$
iii) B(−2, 2) **iv)** $3\sqrt{5}$

16 $x = 2, y = 1$
Line $y = -2x + 5$ is tangent to circle $(x+2)^2 + (y+1)^2 = 20$ at (2, 1)

17 a) Midpoint M(2, 1); gradient $= -\frac{1}{2}$
b) $y = 2x - 3$
c) C(5, 7) A(−1, −5)
d) $20\sqrt{2}$

19 $a = 3$, $b = -2$, $k = 3$

19 $y = \frac{1}{2}x + 2$ $x^2 + (y-2)^2 = 40$

20 (1, 3) (4, 6) **21** A(4a + 3, 3a − 1)

22 $k = 3$
$n = -1, -3, -5 \ldots$
(any negative odd number)
$n = -3$

REVISION 4

Differentiation

1 $x = 36$

2 $t = 4$, $z = 3$ is a minimum point

3 $9x^2 - 4x + 3$ $4t^3 - \dfrac{6}{t^2}$

4 Min $(-12, -\frac{1}{24})$ **5** $y = -6x + 52$

6 $\dfrac{dA}{dt} = -0.08$ m^2/hr
This is rate of change of area when $t = 3$

7 (16, −32) Min **8** $y = -7x + 7$

9 **a)** $f'(3) = -147$ **b)** $-4 < x < 10$

10 SP is $(-2, 5)$ **11** $k = 6$

12 **b)** $x = 3$, $h = 1.5$ SP is Minimum

13 $x > \frac{27}{8}$ **14** Max $(2p, 4p^3)$ 2^{17}

15 **a)** $N = -S + 150$
c) $T = 625$ thousand pounds

16 -5×10^{-5} N/m

17 $y = \frac{1}{4}x + \frac{15}{2}$ $Q(-32, -\frac{1}{2})$

Sample exam paper

1 **a)** $6x + 3$ **b)** $2x^3 - 5x^2 + 5x + 4$

2 **a)** 1/8 **b)** $20x^{\frac{1}{2}}$ **c)** $5 - 3\sqrt{2}$

3 **i)** $y = \dfrac{1}{x-4}$ **ii)** translation by $\begin{pmatrix} 0 \\ 3 \end{pmatrix}$

4 **i)** $6x^2 - \frac{1}{3}x^{-\frac{4}{3}}$ **ii)** $12x + \frac{4}{9}x^{-\frac{7}{3}}$

5 $-1 < x \leqslant 0$, $x \geqslant 1$

6 **i)** $a = 3$, $b = -1$ and $c = -1$
ii) $1 \pm \frac{1}{3}\sqrt{3}$

7 **i)** 1/2 **ii)** $x - 2y = -5$

8 **i)** (1, 4) and (3, 0)
ii) (1, 4) max (3, 0) min

9 **i)** $k = \pm 3\sqrt{2}$ **ii)** 6 units

INDEX

≡ 14

a^0 3
$a^{1/n}$ when n is even 4–5
$a^{1/n}$ when n is odd 4
$a^{m/n}$ 5
a^{-n} 3–4
$a + b\sqrt{x}$, division of an expression by 67–71
addition of polynomials 13
areas of shapes 58–60
$ax^2 + bx + c$ 77–9
  completed square format for
  discriminant of equation $ax^2 + bx + c = 0$ 85–7

base 1
boxing off procedure, areas and 58–61

centroid of triangle 45
circle
  equation of 137–8
  geometrical properties 140–3
  intersection of line with 143–6
  tangent to 138–40
coefficient of a polynomial 12
common factors 18–19
completed square form 31
completing the square 30–4
  for $ax^2 + bx + c$ 77–9
  for $x^2 + bx + c$ 30–3
  solving the equation $ax^2 + bx + c$ 33
conjecture of gradient of curve 91
constant term 12
continuous graph 117
cubic function 12

derivative of a curve 96
  second 128–9, 129–31
difference of two squares 20
  surds and 67
differentiation 96–9, 101–3
  for gradients in calculations 97–8
  notation 96–7
  see also differentiation, applications of
differentiation, applications of 115–36
  increasing and decreasing functions 116–17
  normals 125–7
  optimisation problems 121–5
  rates of change 115–16
  second derivative 128–9, 129–31
  stationary points 117–20, 129–31
  tangents 125–7
discriminant of equation
  $ax^2 + bx + c = 0$ 85–7
  definition 85
  quadratic equations and 112–13
division
  of an expression by $a + b\sqrt{x}$ 67–71
  of indices 2

equations of straight lines 48–53
  alternative forms 50–1
  definition 48
  sketching straight lines from their equations 52

factorisation of polynomials
  further 24–6
  quadratic expression
    $ax^2 + bx + c$ 22–4
    $x^2 + bx + c$ 19–21
  simple 26
  two step 21–2
factorisation of quadratic equations 28–30
function notation 11–12

gradient of a curve 89–107
  definition 89
  formalising the process of finding 104–5
  practical differentiation 96–9, 101–3
  small chords method for estimating 90–3
  tangents and gradients 89–90, 112–13
  using a spreadsheet
    $f(x) + g(x)$ 99–101
    $y = x^n$ 93–5
gradient of line joining two points 46–7
gradient of parallel lines 53
gradient of perpendicular lines 53
gradient of the tangent 89
graphs
  basic transformations 156–7
  commonly occurring 148–50
  continuous 117
  effect of reflections on 158–62
  effect of stretches on 162–6
  effect of transformations on 156–66
  effect of translations on 158–62
  of polynomials given in factorised form 151–6
  quadratic function 72–4
    that can be factorised 74–7
    that cannot be factorised 79–81

identities 14–15
  proving 14
index, definition 1
indices 1–9
  basic rules 1–3
  division 2
  equations with 7–9

indices (*continued*)
evaluating or simplifying expressions involving 6
interpreting a^0 3
interpreting $a^{1/n}$ when n is even 4–5
interpreting $a^{1/n}$ when n is odd 4
interpreting $a^{m/n}$ 5
interpreting a^{-n} 3–4
multiplication 1
negative 8
positive 8
powers of products and quotients 5
raising a power to a power 1
$x^n = k$ 8
inflection, points of 131–3
intercept of the line 49

line of symmetry through graph 72, 74–6

maximum point of curve 80–1, 129, 130, 133
median of triangle 45
minimum point of curve 129, 130, 132, 133
multiplication
of indices 1
of polynomials 13

normals 125–7

parallel lines 53–6
gradient of 53
perpendicular lines 53–6
gradient of 53
points of inflection 131–3
polygons, areas of 58–61
polynomials
algebra 16–17
arithmetic of 13–14
constant term 12
definition 12
degree of 12
factorisation of quadratic expression
$ax^2 + bx + c$ 22–4
$x^2 + bx + c$ 19–21
factorisation of simple 26
further factorisations 24–6
identical 15
identities 14–15
two step factorisations 21–2
positive square root 4
power 1

quadratic equations 27–41
completing the square 30–4
definition 27
equations that are reducible to 108–10
formula for solving 34–5
in solving problems 36–40
linear/quadratic simultaneous equations 110–12
solving by factorisation 28–30
solving simple 27
using the discriminant 112–13
quadratic function 12, 72–88
completed square format for $ax^2 + bx + c$ 77–9
discriminant of the equation
$ax^2 + bx + c = 0$ 85–7
graph of 72–4
that can be factorised 74–7
that cannot be factorised 79–81
linear inequalities 81–2
quadratic inequalities 83–4

rates of change 115–16
rational number, definition 64
rationalising the denominator 66
reflections, effect on graph 158–62
repeated root of an equation 85
root of an equation 28

second derivative 128–9
in determining nature of stationary point 129–31
points of inflection 131–3
shapes, areas of 58–60
simultaneous equations
in finding points of intersection 56–8
one linear, one quadratic 110–12
small chords method for estimating gradient of curve 90–3
solution of an equation 28
stationary points 117–20, 129–31
straight lines, equations of 48–53
alternative forms 50–1
definition 48
sketching 52
stretches, effect on graph 162–6
subtraction of polynomials 13
surd form 64
surds 64–71
division of an expression by $a + b\sqrt{x}$ 67–71
simplifying 64–7

tangent to the curve 89–90, 112–13, 125–7
gradient of 89
transformations, effect on graph 156–66
translations, effect on graph 158–62
triangle
centroid of 45
medians of 45
two points
distance between 43–4
gradient of line joining 46–7
midpoint of line segment joining 43–4

vector notation 42
vertex of curve, definition 72